NATIONAL ACADEMIES — *Sciences Engineering Medicine*

NATIONAL ACADEMIES PRESS
Washington, DC

Automated Research Workflows for Accelerated Discovery

Closing the Knowledge Discovery Loop

Committee on Realizing Opportunities for Advanced and Automated Workflows in Scientific Research

Board on Research Data and Information

Board on Mathematical Sciences and Analytics

Committee on Applied and Theoretical Statistics

Computer Science and Telecommunications Board

Division on Engineering and Physical Sciences

Policy and Global Affairs

Consensus Study Report

NATIONAL ACADEMIES PRESS 500 Fifth Street, NW, Washington, DC 20001

This activity was supported by contracts between the National Academy of Sciences and Schmidt Futures. Any opinions, findings, conclusions, or recommendations expressed in this publication do not necessarily reflect the views of any organization or agency that provided support for the project.

International Standard Book Number-13: 978-0-309-68652-5
International Standard Book Number-10: 0-309-68652-0
Digital Object Identifier: https://doi.org/10.17226/26532

This publication is available from the National Academies Press, 500 Fifth Street, NW, Keck 360, Washington, DC 20001; (800) 624-6242 or (202) 334-3313; http://www.nap.edu.

Copyright 2022 by the National Academy of Sciences. National Academies of Sciences, Engineering, and Medicine and National Academies Press and the graphical logos for each are all trademarks of the National Academy of Sciences. All rights reserved.

Printed in the United States of America.

Suggested citation: National Academies of Sciences, Engineering, and Medicine. 2022. *Automated Research Workflows for Accelerated Discovery: Closing the Knowledge Discovery Loop*. Washington, DC: The National Academies Press. https://doi.org/10.17226/26532.

The **National Academy of Sciences** was established in 1863 by an Act of Congress, signed by President Lincoln, as a private, nongovernmental institution to advise the nation on issues related to science and technology. Members are elected by their peers for outstanding contributions to research. Dr. Marcia McNutt is president.

The **National Academy of Engineering** was established in 1964 under the charter of the National Academy of Sciences to bring the practices of engineering to advising the nation. Members are elected by their peers for extraordinary contributions to engineering. Dr. John L. Anderson is president.

The **National Academy of Medicine** (formerly the Institute of Medicine) was established in 1970 under the charter of the National Academy of Sciences to advise the nation on medical and health issues. Members are elected by their peers for distinguished contributions to medicine and health. Dr. Victor J. Dzau is president.

The three Academies work together as the **National Academies of Sciences, Engineering, and Medicine** to provide independent, objective analysis and advice to the nation and conduct other activities to solve complex problems and inform public policy decisions. The National Academies also encourage education and research, recognize outstanding contributions to knowledge, and increase public understanding in matters of science, engineering, and medicine.

Learn more about the National Academies of Sciences, Engineering, and Medicine at **www.nationalacademies.org**.

Consensus Study Reports published by the National Academies of Sciences, Engineering, and Medicine document the evidence-based consensus on the study's statement of task by an authoring committee of experts. Reports typically include findings, conclusions, and recommendations based on information gathered by the committee and the committee's deliberations. Each report has been subjected to a rigorous and independent peer-review process and it represents the position of the National Academies on the statement of task.

Proceedings published by the National Academies of Sciences, Engineering, and Medicine chronicle the presentations and discussions at a workshop, symposium, or other event convened by the National Academies. The statements and opinions contained in proceedings are those of the participants and are not endorsed by other participants, the planning committee, or the National Academies.

Rapid Expert Consultations published by the National Academies of Sciences, Engineering, and Medicine are authored by subject-matter experts on narrowly focused topics that can be supported by a body of evidence. The discussions contained in rapid expert consultations are considered those of the authors and do not contain policy recommendations. Rapid expert consultations are reviewed by the institution before release.

For information about other products and activities of the National Academies, please visit www.nationalacademies.org/about/whatwedo.

COMMITTEE ON REALIZING OPPORTUNITIES FOR ADVANCED AND AUTOMATED WORKFLOWS IN SCIENTIFIC RESEARCH

DANIEL ATKINS (NAE) (*Chair*), W. K. Kellogg Professor Emeritus of Information and Professor Emeritus of Electrical Engineering and Computer Science, University of Michigan

ILKAY ALTINTAS, Chief Data Science Officer, San Diego Supercomputer Center and Founding Fellow, Halicioglu Data Science Institute, University of California, San Diego

SHREYAS CHOLIA, Group Leader, Usable Software Systems Group, Lawrence Berkeley National Laboratory

MERCÈ CROSAS, Secretary of Open Government, Government of Catalunya

ALFRED HERO, R. Jamison and Betty Williams Professor of Engineering, Department of Electrical Engineering and Computer Science, University of Michigan

REBECCA LAWRENCE, Managing Director, F1000 Research Ltd, London, UK

BRADLEY MALIN (NAM), Accenture Professor of Biomedical Informatics, Biostatistics and Computer Science, Vanderbilt University

LARA MANGRAVITE, President, Sage Bionetworks

BRIAN NOSEK, Executive Director, Center for Open Science*

TAPIO SCHNEIDER, Theodore Y. Wu Professor of Environmental Science and Engineering and Jet Propulsion Laboratory Senior Research Scientist, California Institute of Technology

*Resigned from committee, May 26, 2020

Principal Project Staff

TOM ARRISON, Study Director and Director, Board on Research Data and Information

EMI KAMEYAMA, Program Officer, Board on Research Data and Information

GEORGE STRAWN, Scholar, Board on Research Data and Information

ESTER SZTEIN, Deputy Director, Board on Research Data and Information

OLIVIA TORBERT, Senior Program Assistant, Board on Research Data and Information (until March 2022)

JON EISENBERG, Senior Director, Computer and Telecommunications Board

MICHELLE SCHWALBE, Director, Board on Mathematical Sciences and Analytics

PAULA TARNAPOL WHITACRE, Consultant Writer

BOARD ON RESEARCH DATA AND INFORMATION

SARAH M. NUSSER (*Chair*), Professor, Department of Statistics, Iowa State University
AMY BRAND, Director, MIT Press
BONNIE CARROLL, Retired Founder and Strategic Consultant, Information International Associates, Inc. (CODATA Secretary General)*
STUART I. FELDMAN, Chief Scientist, Schmidt Futures
IAN T. FOSTER, Senior Scientist, Argonne National Laboratory and Distinguished Fellow and the Arthur Holly Compton Distinguished Service Professor of Computer Science, University of Chicago.
RAMANATHAN GUHA, Google Fellow and Vice President, Google, Inc.
JOHN HILDEBRAND (NAS), Regents Professor of Neuroscience, University of Arizona (NAS Foreign Secretary)*
SALLIE A. KELLER (NAE), Distinguished Professor in Biocomplexity, Director of the Social and Decision Analytics Division within the Biocomplexity Institute and Initiative, University of Virginia
MARY LEE KENNEDY, Executive Director, Association of Research Libraries
BAREND MONS, Chair in Biosemantics, Leiden University Medical Center
MICHAEL STEBBINS, President, Science Advisors, LLC

*Denotes ex-officio member

COMPUTER SCIENCE AND TELECOMMUNICATIONS BOARD

LAURA M. HAAS (NAE) (*Chair*), Dean, Manning College of Information & Computer Sciences, University of Massachusetts, Amherst
DAVID E. CULLER (NAE), Professor, Electrical Engineering and Computer Science, University of California, Berkeley
ERIC HORVITZ (NAE), Chief Scientific Officer, Microsoft Research
CHARLES ISBELL, Dean of Computing and John P. Imlay Jr. Chair, Georgia Institute of Technology
ELIZABETH MYNATT, Distinguished Professor and Executive Director, College of Computing, Georgia Institute of Technology
CRAIG PARTRIDGE, Professor and Department Chair, Colorado State University
DANIELA RUS (NAE), Andrew (1956) and Erna Viterbi Professor, Computer Science and Artificial Intelligence Lab, Massachusetts Institute of Technology
FRED B. SCHNEIDER (NAE), Samuel B Eckert Professor of Computer Science, Cornell University
MARGO I. SELTZER (NAE), Canada 150 Research Chair and Cheriton Family Chair, Computer Science, University of British Columbia

NAMBIRAJAN SESHADRI (NAE), Professor of Practice, Electrical and Computer Engineering, University of California, San Diego
MOSHE Y. VARDI (NAS/NAE), George Distinguished Service Professor in Computational Engineering, Rice University

BOARD ON MATHEMATICAL SCIENCES AND ANALYTICS

MARK L. GREEN (*Chair*), Distinguished Research Professor in the Department of Mathematics, University of California, Los Angeles
HÉLÈNE BARCELO, Deputy Director, Mathematical Sciences Research Institute
BONNIE BERGER (NAS), Simons Professor of Mathematics, Department of Mathematics, Computer Science and AI Lab, Massachusetts Institute of Technology
RUSSEL E. CAFLISCH (NAS), Director of the Courant Institute, Professor in the Mathematics Department, New York University
DAVID CHU, Adjunct Staff, Institute for Defense Analyses
DUANE COOPER, Associate Professor of Mathematics, Morehouse College
JAMES (JIM) CURRY, Professor, Applied Mathematics Department, University of Colorado Boulder
RONALD FRICKER, Professor of Statistics and Associate Dean, College of Science, RR
TRACHETTE JACKSON, Professor, University of Michigan
LYDIA KAVRAKI (NAM), Noah Harding Professor of Computer Science, Bioengineering, Electrical and Computer Engineering, and Mechanical Engineering, Rice University
TAMARA KOLDA, Distinguished Member of Technical Staff, Sandia National Laboratories
PETER KOUMOUTSAKOS, Professor for Computational Science, ETH Zurich
RACHEL KUSKE, Professor of Mathematics and Department Chair, Georgia Institute of Technology
YANN A. LECUN (NAS/NAE), Professor, Courant Institute of Mathematical Sciences and Center for Data Science, New York University
JILL C. PIPHER, Vice President for Research and Elisha Benjamin Andrews Professor of Mathematics, Department of Mathematic, Brown University
YORAM SINGER, Chief AI Scientist, WorldQuant
TATIANA TORO, Craig McKibben & Sarah Merner Professor of Mathematics, University of Washington
LANCE WALLER, Rollins Professor and Chair, Department of Biostatistics and Bioinformatics, Rollins School of Public Health, Emory University
AMIE WILKINSON, Professor of Mathematics, University of Chicago
KAREN E. WILLCOX, Director, Institute for Computational Engineering and Sciences, Professor of Aerospace Engineering and Engineering Mechanics, University of Texas at Austin

Preface

Exponential improvement in computing and communication technology has revolutionized the conduct of research beginning in 1946 when ENIAC began computing ballistic trajectories at a 3,000× speedup over manual methods. The tools for research have evolved from a few warehouse-size automated calculators to a distributed global cyberinfrastructure supporting observation, control, calculation, information access, publishing, and collaboration. The growing capacity and capability of computers and data repositories enables multiscale, multiscience analysis, modeling, and prediction at ever-increasing temporal and spatial resolution. The concept of e-science or research collaboratory laboratories without walls, reducing barriers of participation, time, and distance (both physical and disciplinary), is pervasive. These concepts have taken us beyond just automating traditional research methods to the possibility of tackling new problems in new ways.

While this stepwise revolution is enabled by hardware and software engineering advances, their meaningful use in research requires intentional nurturing. As the technology advances, early adopters find ways to use it for competitive advantage, and others take note. But broad adoption of new tools and methods, and especially supporting the required cyberinfrastructure (e-research infrastructure), the human resources, and changes in the community of practice require community-driven leadership from research sponsors, the researchers themselves, and their institutions. A combination of bottom-up, community-initiated workshops together with institutionally commissioned studies strives to articulate new opportunities and how to realize them. It is hoped that these motivate research sponsors and performers, scholarly publishers, professional societies, archivers, indexers, and other stakeholders to work together to bring the new capabilities to scale and into meaningful applications. Past examples of such cooperation include the National Science Foundation (NSF) and Department of Energy supercomputer centers, NSFNet, the NSF-ARPA (Advanced Research Projects Agency)-NIH (National Institutes of Health) digital library initiatives, and many more recent initiatives around cyberinfrastructure-enhanced research and data science.

In this spirit, this National Academies of Sciences, Engineering, and Medicine (National Academies) consensus study, commissioned by Schmidt Sciences, is intended as a contribution to a next step in the transformative application of computing to scientific discovery. Leveraging what has come before, researchers are now incorporating artificial intelligence (AI) and the automation of scientific instruments into the research workflow. The focus of this report is not only the use of methods of AI and machine learning (ML) as a component in a workflow, but also the use of these methods to *design* experiments and to *automatically control* them. The goal is to use them in an iterative loop by using experiments or observation data to test and learn about a model, and then to use AI and ML methods to generate the design for the next data collection. This closed loop iterates and, in examples we present in the report, accelerates discovery by orders of magnitude. We refer to these as automated research workflows (ARWs).

Realizing the potential of ARWs is a complex mix of technology, funding, policy, regulation, ethics, education, reward structures, and the overall sociology of varied research communities of practice. ARWs offer benefits that go beyond accelerating exploration, including enhanced capture of provenance, integrity, reproducibility, and dissemination. But achieving the benefit of ARWs depends upon progress in addressing the same fundamental issues that persist in most other explorations of the next big thing in cyberinfrastructure-enhanced computing. These include long-term sustainability of cyberinfrastructure (computing, networking, and now critically, data and programs); reducing barriers and increasing incentives for interdisciplinary collaboration; addressing security challenges; and educating current practitioners and students to design and responsibly use ARWs.

This report is built upon the contributions of many people: 9 dedicated members of the study committee, the 23 agenda speakers and many other participants in a workshop on March 16–17, 2020, and National Academies staff. This report, like many things since March 2020, has been produced under unusual circumstances. We converted the workshop from onsite in Washington to distance independent on 3 days' notice with surprisingly good results. We engaged 120 people for 2 days over 9 time zones and gathered more input than usual through real-time recordings and the mining of chat streams. The subsequent report was then produced through a series of distributed meetings. Although a report like this can be successfully produced in a never-in-the-same-room way, we hope that post-pandemic studies will include some use of trust-building face-to-face meetings and social events.

Sixty years ago, I programmed my first computer, an IBM 1620, and since then have had the enormous privilege to participate in the astonishing computer revolution. As part of that, I have had the honor of chairing several study committees of extraordinary people, such as this committee, to help describe emerging opportunities and challenges for transforming research or learning. My thanks to all who have made this possible. I conclude with the hope that this report will indeed be a contribution to empowering transformative research and its application to a better world.

Daniel E. Atkins
Chair, Committee on Realizing Opportunities for Advanced
and Automated Workflows in Scientific Research

Acknowledgments

This Consensus Study Report was reviewed in draft form by individuals chosen for their diverse perspectives and technical expertise. The purpose of this independent review is to provide candid and critical comments that will assist the National Academies of Sciences, Engineering, and Medicine in making each published report as sound as possible and to ensure that it meets the institutional standards for quality, objectivity, evidence, and responsiveness to the study charge. The review comments and draft manuscript remain confidential to protect the integrity of the deliberative process.

We thank the following individuals for their review of this report: Carole Goble, University of Manchester; Brooks Hanson, American Geophysical Union; Daniel Katz, University of Illinois; Gary King, Harvard University; Robert Murphy, Carnegie Mellon University; Kristin Persson, Lawrence Berkeley National Laboratory; Beth Plale, Indiana University; and Margo Seltzer, University of British Columbia.

Although the reviewers listed above provided many constructive comments and suggestions, they were not asked to endorse the conclusions or recommendations of this report, nor did they see the final draft before its release. The review of this report was overseen by Philip Neches, California Institute of Technology. He was responsible for making certain that an independent examination of the report was carried out in accordance with the standards of the National Academies and that all review comments were carefully considered. Responsibility for the final content rests entirely with the authoring committee and the National Academies.

Contents

SUMMARY		1
1	**INTRODUCTION**	13
	Work of the Committee, 14	
	Scope of the Study and Organization of This Report, 15	
2	**CONTEXT FOR AUTOMATED RESEARCH WORKFLOWS**	19
	The Opportunity, 19	
	Building Automated Research Workflows: Current State of the Art, 24	
	Implementing Automated Research Workflows: A Changing Scientific Paradigm, 28	
	Policy and Industry Context for Automated Research Workflows, 32	
3	**AUTOMATED RESEARCH WORKFLOWS IN ACTION**	37
	Physical Sciences, 40	
	Biomedical Science, 46	
	Health and Environment, 49	
	Digital Humanities, 56	
	Social and Behavioral Sciences, 58	
4	**AUTOMATIC RESEARCH WORKFLOWS AND IMPLICATIONS FOR ADVANCING RESEARCH INTEGRITY, REPRODUCIBILITY, AND DISSEMINATION**	63
	Integrity, 63	
	Reproducibility and Replicability, 66	
	Dissemination, 67	

5	**OVERCOMING BARRIERS TO WIDER USE OF AUTOMATED RESEARCH WORKFLOWS**	**73**
	Reimagining Incentives, 73	
	Overcoming Barriers in the Research Culture, 75	
	Closing Education Gaps, 76	
	Ensuring Sustainability, 78	
	Managing Issues Outside of Research that Affect ARWs, 84	
6	**CONCLUSION**	**89**
	Findings and Recommendations, 90	

BIBLIOGRAPHY 95

APPENDIXES

A	Workshop Agenda	109
B	Committee Biosketches	113

Abbreviations and Acronyms

ADHO	Alliance of Digital Humanities Organizations
AI	artificial intelligence
ARW	automated research workflow
ATLAS	A Toroidal LHC ApparatuS
CSTCloud	China Science and Technology Cloud
CWL	Common Workflow Language
DARPA	Defense Advanced Research Projects Agency
DOD	Department of Defense
DOE	Department of Energy
DUA	data use agreement
EC	European Commission
EOSC	European Open Science Cloud
EU	European Union
FAIR	findable, accessible, interoperable, and reusable
GFLOPS	billion floating point operations per second
HPC	high-performance computing
HTC	high-throughput computing
iREDS	Institutional Re-Engineering of Ethical Discourse in STEM

LHC	Large Hadron Collider
ML	machine learning
NASEM	National Academies of Sciences, Engineering, and Medicine
NIH	National Institutes of Health
NSF	National Science Foundation
OSTP	Office of Science and Technology Policy
RDA	Research Data Alliance
SDSS	Sloan Digital Sky Survey
UKRI	United Kingdom Research and Innovation

Summary

Human society increasingly looks to the research enterprise to help address a wide range of complex and interrelated challenges, including climate change and other environmental issues, emerging infectious diseases and other threats to health, food insecurity, and economic and social disparities. In addition, researchers are working to expand our understanding of fundamental scientific questions, including the origins of the universe, the nature of matter, and the evolution of language. The expanding capability, capacity, integration, and ubiquity of advanced information technology supporting research, also known as cyberinfrastructure, provide ever more opportunity for discovery through modeling, simulation, and prediction, now aided by the practical application of machine learning (ML) and artificial intelligence (AI). Accelerating progress depends on leveraging the exponential growth in the amount and variety of data available through the development and use of sophisticated computational approaches. Computation and automation of data acquisition allow automated systems to analyze data and extract knowledge, to create predictive models, and to use those models to guide the acquisition of additional data, closing and automating the loop in a typical research workflow.

The committee uses the term automated research workflow (ARW) to describe scientific research processes emerging across a variety of disciplines and fields. ARWs integrate computation, laboratory automation, and tools from AI in the performance of tasks that make up the research process, such as designing experiments, observations, and simulations; collecting and analyzing data; and learning from the results to inform further experiments, observations, and simulations. Although the specific tools and resources used and the tasks performed vary by field, the common goal of researchers implementing ARWs is to accelerate scientific knowledge generation, potentially by orders of magnitude, while achieving greater control and reproducibility in the scientific process.

The tools and techniques being developed under the large umbrella of ARWs promise to transform the centuries-old serial method of research investigation into processes in which thousands or even millions of simulations or experiments are iterated rapidly in closed loops, with the analysis of data and even the design of experiments or controlled observations being assisted by ML or optimization techniques. Simultaneously, ARWs provide a way to satisfy pressing demands across fields to increase interoperability, reproducibility, replicability, and trustworthiness by better tracking results, recording data, establishing provenance, and creating more consistent metadata than even the most dedicated researchers can provide themselves.

To explore the benefits and challenges, as well as to suggest opportunities to move forward, the National Academies of Sciences, Engineering, and Medicine's Board on Research Data and Information, in collaboration with the Board on Mathematical Sciences and Analytics and the Computer Science and Telecommunications Board, launched a study aimed at examining current efforts to develop advanced and automated workflows to accelerate research progress. A committee of nine members undertook the study, with support provided by Schmidt Futures. To accomplish its task, committee members held a public workshop in March 2020 and numerous additional discussions, conducted a review of the literature, and drew on their own expertise in specific domains and cross-disciplinary areas such as open science and digital transformation.

Although impressive strides are being made to apply ARWs in a variety of fields, there are also significant barriers to progress. Development and widespread adoption of ARWs require new skills, supportive funding mechanisms, and shifts in culture. Concerns about the role of humans in the discovery loop, privacy of data, and the impact on current incentive systems need to be addressed. What unforeseen technical and ethical issues may arise? Who "owns" the data and discoveries that are produced by automated and distributed systems? How should researchers evolve their practices to reap the benefits of automation while not losing the serendipity of human inspiration and creativity? What goals are best achieved by human scientists (such as invention of new techniques) and which are better left to automation (such as driving data collection to optimize models)?

Given the newness and rapidly evolving nature of this topic, the committee's findings and recommendations are necessarily broad and future oriented. Fully realized ARWs are not common at present, and so the study examines how and where progress is being made in areas such as advanced computation, use of workflow management systems, laboratory automation, and the use of AI both as a workflow component as well as in directing the "outer loop" of the research process. This report constitutes an initial effort to create awareness, momentum, and synergies to help realize the potential of ARWs in scientific discovery. The committee hopes that this report will stimulate further discussion, transformations, and, most important, investments and meaningful use.

FINDING A: ACCELERATING DISCOVERY

In many disciplines, the emergence of automated research workflows (ARWs), built upon contemporary cyberinfrastructure, is demonstrating the potential to vastly increase the speed and efficiency of a range of research activities. These include designing and conducting experiments, analyzing data, and observing natural phenomena. These improvements can be realized at scale by implementing infrastructure and practices that facilitate the application of artificial intelligence and machine learning and related technologies to research. Realizing the potential of ARWs could accelerate the pace of scientific discovery by orders of magnitude and thereby expand the research enterprise's contribution to society.

Developments in multiple domains are beginning to deliver on this potential. Below are some examples:

- In materials science, research groups are building systems in which a combination of laboratory automation and ML are cutting the time required for synthesis and testing of materials from 9 months to 5 days (Service, 2019).
- In particle physics, new approaches to drawing inferences can be combined in workflows that implement various inference algorithms. The new approaches hold the possibility of significantly advancing the productivity of research by allowing experiments to achieve, for example, a given sensitivity using half the data (Cranmer, 2020).
- In drug discovery, an active learning algorithm identified 57 percent of the active compounds by performing 2.5 percent of the possible experiments, compared with 20 percent identified through a traditional approach of building a model for each target (Kangas et al., 2014).
- Researchers working in biochemistry are using robotics and data science to automate high-throughput synthesis and screening (Cernak, 2020).
- Astronomers are using ML and increasingly fine controls on telescopes to automate target selection so that observations are optimally informative given the observational constraints and scientific objectives (Szalay, 2020).
- In climate science, the generation of high-resolution local simulations to inform lower-resolution global climate models about important small-scale processes can be automated, closing the loop of generating computational experiments and informing a global model with them (Schneider et al., 2017a).
- In digital humanities, scholars are using deep neural annotation networks to tackle the complexities of compiling information from huge volumes of words and across multitudes of languages over the centuries to see patterns in how ideas have spread and changed over time, and to understand the development of human thought (Crane, 2020).

- Researchers in the social and behavioral sciences are using new data resources and advanced analytics to better understand and address a range of pressing problems, including poverty alleviation and strengthening the delivery of public services in cities (O'Brien et al., 2017).[1]

FINDING B: ADDITIONAL BENEFITS

In addition to increasing the speed and efficiency of research, the effective development and implementation of the technical and human infrastructure for automated research workflows (ARWs) will contribute to strengthening the research process in other ways. For example, the greater transparency and repeatability made possible by automating and capturing specific steps in the research process—advances that underlie the development of ARWs—can foster reproducibility, replicability, and responsibility in research. Adoption of common and interoperable tools and platforms—which could be accelerated by the advance of ARWs but depends on other developments as well—can facilitate international and interdisciplinary research collaboration. Broader access to research workflows and results and the enhanced ability to uncover and correct errors can contribute to greater confidence in research findings and the research enterprise and reduce redundancy among research efforts. In addition, incorporating emerging principles and guidelines for responsible artificial intelligence and machine learning advocated by various organizations, such as building in human review of algorithms, uncovering and addressing bias, and supporting transparency and reproducibility, will also help to secure the benefits of ARWs.

ARWs can help strengthen transparency, rigor, and reliability in research in several ways. These include the use of scientific workflow engines, software that provides a formalization of the computational analysis pipeline, as well as platforms to facilitate data flows and data management pipelines (ODSC Community, 2021). These tools provide a structured and repeatable process to automate complex ARWs at scale.

An additional trend offering an on-ramp to ARWs is the use of interactive computational laboratory notebooks, such as Jupyter, that allow researchers to capture a set of discrete analysis steps and track them with a single user interface. There were about 9.7 million Jupyter notebooks stored on GitHub when this was written in November 2020, with the number growing by about 8,500 per day (Project Jupyter, 2020). Increasingly sophisticated platforms, such as Google's AI Platform and RedHat Openshift AI/ML workflows that integrate Jupyter Notebooks or similar interactive interfaces with scalable compute and data resources, are enabling complex, iterative AI/ML pipelines (Google Cloud, 2020; RedHat Openshift, 2020).

---

[1] For more information, see also https://opportunityinsights.org.

RECOMMENDATION 1: Design Principles

Organizations that fund, perform, and disseminate research, along with scientific societies, should support and enable automated research workflows (ARWs) that embody the following design principles:

- ARWs and the systems, tools, and platforms that comprise them should facilitate openness, reproducibility, and transparency.
- ARWs should facilitate the effective use of artificial intelligence (AI) and machine learning (ML) as research tools and incorporate principles of responsible AI and ML to mitigate the risks from various human and technological deficiencies, such as confirmation and sampling biases, inappropriate application of statistics, and challenges to interpretability of results and quantification of confidence and uncertainties when drawing inferences from ML analyses.
- The associated research objects (data, code, even entire workflows) for ARWs should be FAIR (findable, accessible, interoperable, and reusable), not only by humans but also by machines, to facilitate automated reuse and collaboration.
- ARWs should prioritize reuse and sustainability of existing tools and systems when possible and appropriate, reducing costly duplication efforts and facilitating the extension of capabilities through integration or federation of systems, and agreement on standards. Designs should allow for specialization into specific domains, but avoid unnecessary rebuilding.
- While proprietary services and components can enhance the utility of ARWs, key ARW infrastructure should be controlled by and be accessible to the research community itself, with the community developing standards and practices to facilitate this.

This list of principles constitutes a vision for future ARW development that emerged from the study. They support and expand upon the recommendations of several National Academies reports that underline the importance of openness and reproducibility in research (NASEM, 2017, 2018b, 2019b). Although ARWs can be implemented without open research, open data, code, and workflows can speed their adoption and enhance their impact.

As these principles and other recommendations demonstrate, the focus of the committee's effort went beyond the use of AI as a component in a workflow to the use of AI methods to *design* experiments and to *automatically control* them. We offer our findings and recommendations related to design principles, infrastructure sustainability, human resources, culture and incentives, and privacy protection as contributions to the next step in the transformative application of computing to scientific discovery.

There was a great deal of discussion at the workshop about mitigating various risks to the effective utilization of ARWs that can arise from technological and human deficiencies. Workflows do not guarantee transparency or, more broadly, integrity. An overreliance on naive ML approaches can lead to severe biases and lack of replicability. An example is *p*-hacking which occurs when successive ML algorithms are applied to data and only the algorithm with best performance is reported (Nugent, 2020).

Examples discussed at the March 2020 workshop point to the benefits of building ARWs on common lower-level, interoperable infrastructure to the extent possible as opposed to being project specific. The Common Workflow Language (CWL) is a good example of a standard aimed at facilitating interoperability (CWL, 2020).

FINDING C: RESEARCH ENTERPRISE

Realizing the potential of automated research workflows (ARWs) will require modification of the research enterprise, including sustainable funding for the necessary hardware, software, and human resources, educating the scientific workforce, reporting and sharing research results, and structuring researcher rewards and incentives. Multidisciplinary, multirole collaboration is essential to realize the potential of ARWs.

Because solutions to the biggest problems of our time are complex, they require end-to-end workflows for integrated management of many technical steps in addition to extensive knowledge of the application domain. These integrated steps require expertise from multidisciplinary team members to collaborate on (1) methods to manage, integrate, and interpret "big" data; (2) modeling and simulation tools executing on scalable computing platforms; (3) methods and interfaces for domain-specific analysis, communication, and visualization of results; and (4) technologies to make the process FAIR, that is, portable, transparent, repeatable, and reproducible.[2] Such a multidisciplinary collaboration to solve a problem is different from the way an individual conducts a scientific study, shifting the paradigm from individual science to team science.

In addition to creating a more supportive environment for the development and implementation of ARWs, the actions identified here will support more effective use of AI/ML and advanced computation in research more broadly.

RECOMMENDATION 2: Infrastructure, Code, and Data Sustainability

Research funders, working with other stakeholders such as societies, research institutions, and publishers, should place greater priority on approaches to ensuring the creation and sustainability of key systems, tools,

---

[2] GO FAIR, https://www.go-fair.org/. Accessed August 21, 2020.

platforms, and data archives for automated research workflows (ARWs). Priorities include

- **Funding support for efforts by research institutions and societies to link disciplines so they can share and benefit from the expertise in statistics, machine learning or data science, and engineering and computer science that is required to build and maintain sustainable infrastructure for ARWs.**
- **Funders and research communities structuring funding for cyberinfrastructure projects such as large scientific instruments so as to maximize the potential for innovation in ARWs and the reuse of data and other outputs.**
- **Funders and research communities supporting open data standards and open interfaces for scientific instruments.**
- **Funders and research institutions enabling reuse, reproducibility, and long-term sharing of FAIR data and software resources through support of repositories that make archival and updated versions of these resources available within and across disciplines, and providing approaches to sustain those repositories.**
- **Publishers updating their data-sharing requirements by directly associating articles to data in FAIR repositories.**

Although providing sustained support for the development and operation of shared cyberinfrastructure across multiple disciplines has been challenging in the United States, there are numerous historical success stories such as the National Science Foundation (NSF) National Supercomputer Centers, the development of GenBank and other digital data resources in the life sciences, and NSF's advanced cyberinfrastructure program. Current and recent U.S. programs aimed at establishing shared resources include Harnessing the Data Revolution (NSF, 2020b), the National Institutes of Health (NIH) Data Commons Consortium (NIH, 2018), and a series of efforts to advance strategic computing and related technologies across agencies under the auspices of the Networking and Information Technology Research and Development Program[3] and its predecessors. Currently under way is the Global Open Research Commons Interest Group[4] under the Research Data Alliance (RDA) (Jones, 2021), led by some members of the now terminated Data Commons Consortium. It plans to address the issues related to avoiding silos and agreeing on standards to build globally oriented data and science environments.

On January 1, 2021, Congress passed the National Artificial Intelligence Initiative Act.[5] The act's language reflects many of the issues raised in this report.

---

[3] For more information, see https://www.nitrd.gov/.
[4] For more information, see https://www.rd-alliance.org/groups/global-open-research-commons-ig.
[5] 15 U.S.C. Chapter 119, https://uscode.house.gov/view.xhtml?path=/prelim@title15/chapter119&edition=prelim.

The 2021 federal budget included $868 million (a 76 percent increase) to NSF for AI-related grants and interdisciplinary research initiatives; $125 million to the U.S. Department of Energy's Office of Science on AI research; and $50 million to NIH for research on chronic disease using AI and related approaches (CRS, 2020). For example, NIH is planning to invest $23 million per year over 7 years to support Artificial Intelligence for Biomedical Excellence, which will generate new biomedically relevant data sets amenable to ML (NIH, 2020).

Advancing AI as a research tool and building complementary data stewardship capacities are a focus internationally. In 2020, United Kingdom Research and Innovation, the major public funder of research in the UK, targeted computational and e-infrastructure as a priority theme for research infrastructure investment. The Alan Turing Institute, the UK's national institute for data science and AI, has been strategically investing in key areas as part of a shared vision for the future of AI and data science in the UK. The European Open Science Cloud[6] is still in development but is designed to be a new shared infrastructure with long-term investment to provide access to data repositories as well as resources such as cloud services, high-performance computing, and data analysis tools (EOSC, 2020). And in China, the CSTCloud (China Science and Technology Cloud) is emerging as a national infrastructure of data and computation for accelerating science discovery (CSTCloud, 2020).

Given the significant investments that governments and other research funders are making in data-driven science, it makes sense to leverage these investments across borders and domains to the extent possible. Goals of enhanced international collaboration would include facilitating access to tools and resources and ensuring the interoperability of national implementations. Distributed international efforts such as GO FAIR, RDA, the Research Data Framework, CODATA, FORCE11, and the Research Software Alliance are working to develop standards and approaches to facilitate research data management and sharing and FAIR data and software.

RECOMMENDATION 3: Human Resources
Research funders, higher education, research institutions, and scientific and professional societies should support the development and implementation of educational programs and career pathways aimed at building the workforce needed to develop and utilize automated research workflows (ARWs), including the creation of career tracks that support ARW capabilities. Examples of what is needed include

- **Programs that foster integration of domain expertise with data science and software engineering skills.**
- **Programs that inculcate data literacy and computational analytical skills in all areas of research.**

---

[6] See https://ec.europa.eu/info/research-and-innovation/strategy/goals-research-and-innovation-policy/open-science/european-open-science-cloud-eosc en.

- Developing the human resources needed to build, maintain, and operate ARW hardware and software, including hardware and software engineers who build, maintain, and operate automated laboratories and the software needed to learn from data and to design experiments.
- Fostering collaborative research that aims at developing and using ARWs and that facilitates sharing workflows, code, data, and data products in ways that respect and protect privacy considerations.

Discussion at the March 2020 workshop revealed a number of education and training needs and challenges to effective development and implementation of ARWs that are common across multiple domains. For example, a need for more researchers who combine domain knowledge and data science or software development expertise was expressed in just about all use case discussions. Knowledge of mathematical and computational methods for designing experiments or controlling observations automatically and for learning from data is becoming essential for researchers (NASEM, 2018a, 2020b).

RECOMMENDATION 4: Culture and Incentives
Research funders, research institutions, and disciplines should work to create an automated research workflow (ARW)-friendly culture by making changes in incentive and reward structures aimed at encouraging behaviors that are central to realizing the potential of ARWs. These include

- **Encouraging team science and multidisciplinary teams.**
- **Using funding support and provisions for data management plans to encourage development and curation of FAIR, responsible, and good-quality data resources.**
- **Developing, improving, and sharing software resources.**
- **Reporting reproducible results.**
- **Helping others adopt ARW practices.**
- **Pursuing international collaboration when possible in order to accelerate progress toward implementing the above changes at scale.**

Misalignment of the incentives and priorities of researchers, research institutions, and research funders with the actions and efforts needed to effectively develop and implement ARWs was a major theme of the March 2020 workshop discussion and manifested itself in a variety of ways. Indeed, the rewards and incentives built into the research enterprise as it exists today and their negative impact on efforts to modernize research through greater transparency and rigor go beyond ARWs and have been discussed in several recent National Academies reports (NASEM, 2017, 2018b, 2019b, 2020a). Current conditions are not conducive to creating research environments that encourage transparency, sharing, the formation of interdisciplinary teams, and other behaviors that would boost the effectiveness of ARWs.

The COVID-19 experience illustrates that the prerequisites for automated and rapid research responses to public health threats are the same as those needed to effectively utilize ARWs in general (RDA, 2020). These include common data models, approaches to mitigate bias in data collection and analysis, and support for the infrastructure necessary to facilitate near-real-time scientific investigation and dissemination of findings that lead to guidance on how best to mitigate public health threats. Currently, there are strong incentives for launching rapid research responses, but weak incentives for sharing the outputs and ensuring that such responses are rigorous and reliable.

A shortage of shared domain resources being used, particularly well-characterized FAIR data and related infrastructure such as repositories and active curated services, is apparent in drug discovery, the digital humanities, and other fields. These resources are ultimately critical to the implementation of ARWs, because data are needed to develop and train ML algorithms, which in turn enable the development of closed-loop systems.

The European Union has prioritized open research data. Recent cost-benefit analysis estimated that by not having FAIR data, Europe incurs an opportunity cost of about $10 billion per year in categories such as additional time spent on research, higher storage costs, higher license costs, and higher rates of retracted research findings (EC, 2019). The EU also increased its investments in AI by 70 percent from 2018 to 2020, to about €1.5 billion (EC, 2021a).

Recognizing the overall importance of FAIR principles, their most salient point in the context of ARWs is to make research products such as the data, software, and workflows machine actionable. In addition to FAIR, data quality and data privacy (in some cases) are of great importance, and the FAIR principles do provide for authorization to access private data.

FINDING D: LEGAL AND POLICY ISSUES

In addition to barriers to progress that exist within the research process itself, there are legal and policy issues that affect implementation of automated research workflows in specific domains that will require international multistakeholder efforts to address.

Data collected for use in ARWs will increasingly include data generated outside of a traditional research setting, such as personal health data collected from wearables and medical visits or behavioral data collected online from social media. Use of such data is subject to additional challenges—catalyzed by documented cases of misuse—including public mistrust, institutional policies, and government regulation designed to protect personal data privacy. ARWs will have to be designed to comply with these terms and provide transparency in results from data use. For example, in response to concerns about the security and use of personal data, exacerbated by well-publicized examples of data breaches and misuse, policy makers and public interest groups have pushed to allow individuals to have greater control over the use, storage, and reuse of their data. Prime examples are the European Union's General Data Protection Regulations and the California Consumer Privacy Act. ARWs will need to comply with such policies.

The development and implementation of ARWs is also impacted by broader issues raised by the growing use of AI in a variety of policy-making and decision-making contexts. It will be a challenge to maintain transparency and trust in outcomes with significant real-life consequences when an algorithm determines those outcomes (Stoyanovich et al., 2020a).

RECOMMENDATION 5: Preserving Privacy

Research enterprise funders, performers, publishers, and beneficiaries should work with governments, data privacy experts, and other entities to address the legal, policy, and associated technical barriers to implementing automated research workflows in specific domains. They should explore solutions to make the outputs available through privacy-preserving algorithms and federated learning approaches to using data.

Privacy, ethics, and similar socially based topics will likely emerge as (a) more data are added into the workflow cycle and (b) the questions being asked are modified. Building mechanisms into ARWs that recognize and are sensitive to authorized data use issues is new territory to explore and develop, for example, to be able to distinguish between permitted and prohibited queries (Kusnezov, 2020). Ideas proposed at the workshop included embedding compliance in the design of the software for open research data services and standards for the architecture of the sharing and access system (Burgelman, 2020). Examples of current work on privacy-preserving approaches for social sciences include how to reduce privacy loss when dealing with small sample sizes (Chetty and Friedman, 2019) and development of a mathematical framework to quantify and manage privacy risks (Wood et al., 2018).

1

Introduction

The needs and demands placed on science to address a range of urgent problems are growing. The world is faced with complex, interrelated challenges in which the way forward lies hidden or dispersed across disciplines and organizations. Treatment for and immunization against COVID-19 is the most immediate example at the time of this report, but so, too, is the push in other disease areas such as cancers and Alzheimer's disease, as well as in climate change, natural disaster prevention and mitigation (earthquake risk assessment, hurricane forecasting), agriculture (feeding a growing world population with finite resources), and other critical areas.

For centuries, scientific research has progressed through iteration of a workflow built on experimentation or observation and analysis of the resulting data. While computers and automation technologies have played a central role in research workflows for decades to acquire, process, and analyze data, these same computing and automation technologies can now also control the acquisition of data, for example, through the design of new experiments or decision making about new observations. The committee uses the term automated research workflows (ARWs) to describe scientific research processes that are emerging across a variety of disciplines and fields. ARWs integrate computation, laboratory automation, and tools from artificial intelligence in the performance of tasks that make up the research process, such as designing experiments, observations, and simulations; collecting and analyzing data; and learning from the results to inform further experiments, observations, and simulations. While the specific tools and resources used and the tasks performed vary by field, the common goal of researchers implementing ARWs is to accelerate scientific knowledge generation, potentially by orders of magnitude, while achieving greater control and reproducibility in the scientific process. This enhanced capability, in turn, is enabling researchers to address qualitatively new questions and collaborate more effectively. Artificial intelligence (AI) and machine learning (ML) techniques play an increasingly important role in ARWs, from uses in data exploration and analysis to the driving and directing of the larger research process as a closed-loop system where AI and ML analyses of results direct the next cycle of experimental design and planning. The committee believes that ARWs constitute the next significant advance in the ongoing revolution in scientific research driven by advances in information technology and associated hardware infrastructure.

Although the design of such ARWs remains in the hands of humans, execution can be automated and accelerated. ARWs can help manage and exploit the exponentially expanding amount and availability of data. Within these data may well lie the solutions to many problems the world faces today, as well as to problems that we will confront tomorrow. Without technological assistance and automation, it would be impossible for humans to review, much less use, the enormous data resources that may prove pivotal to discovery. ARWs can increase the speed and quality of discovery. At the same time, ARWs provide a way to satisfy pressing demands across fields to increase interoperability, reproducibility, replicability, and trustworthiness by better tracking results, recording data, establishing provenance, and creating more consistent metadata than even the most dedicated researchers can provide by themselves.

Thus, ARWs can also support more transparent and reliable science. A growing number of research funders are encouraging or requiring that the data, methods (including analytical code), and other artifacts underlying the work that they support be openly available. Wide implementation of ARWs could encourage researchers to make more of their research data findable, accessible, interoperable, and reusable (FAIR) and facilitate data and software reuse and sharing in trusted repositories. Although open and FAIR research outputs are not a prerequisite to the use of ARWs, they are highly desirable and complementary.

WORK OF THE COMMITTEE

In 2019, the National Academies of Sciences, Engineering, and Medicine's Board on Research Data and Information, in collaboration with the Board on Mathematical Sciences and Analytics and the Computer Science and Telecommunications Board, launched a study aimed at examining current efforts to develop advanced and automated workflows to accelerate research progress, including wider use of artificial intelligence (see Box 1-1 for the committee's statement of task). An expert committee undertook the study with support from Schmidt Futures. To accomplish its task, the committee held an initial meeting on August 13, 2019, in Washington, DC, to organize the study process, identify information needs, and develop the agenda and identify participants for a public workshop. As the primary information-gathering

> **BOX 1-1**
> **Committee Statement of Task**
>
> An ad hoc committee of the National Academies of Sciences, Engineering, and Medicine will conduct a study that examines current efforts to develop advanced and automated workflows for scientific research. The study will also identify promising research approaches to accelerating progress in the effectiveness and utilization of workflow systems and tools. The committee's primary information gathering will consist of a workshop that examines the status of research workflows in several example fields, key barriers and enablers, and emerging opportunities. The workshop will explore the role of open science, in the form of broad access to research articles, data, and analytical code, and other enabling factors. Based on insights from the workshop, a review of the literature, and other inputs, the committee will produce a consensus report that identifies research needs and priorities in the use of advanced and automated workflows for scientific research.

mechanism, a 2-day virtual workshop, "Opportunities for Accelerating Scientific Discovery: Realizing the Potential of Advanced and Automated Workflows," was held March 16–17, 2020. The committee defined the subtopics for the study and identified more than 25 outstanding experts to participate in the workshop. Presentation and discussion topics included research use cases, mathematical and algorithmic barriers, trajectories for supporting tools and systems, standards and social context, policy and educational implications, communities and sustainable funding, and transparency and accountability (see Appendix A for the full agenda and the National Academies' website for copies of the presentations).[1] More than 20 virtual committee meetings using collaborative authoring tools were then held to discuss, draft, and finalize this consensus report.

SCOPE OF THE STUDY AND ORGANIZATION OF THIS REPORT

In its deliberations, the committee recognized the multifaceted nature of its charge. Different disciplines of research have very different practices relative to ARWs—in terms of specific tools and platforms and, more generally, propensity to incorporate workflows into their processes in the first place. In addition, several lines of thought that emerged from the March 2020 workshop are germane not just to the task at hand, but more broadly across the scientific enterprise. These themes are (1) breaking down academic silos, (2) providing incentives for greater collaboration among researchers, (3) ensuring greater interoperability across technologies, (4) sharing of a broader range of research outputs, and (5) striking an appropriate

---

[1] Copies of the speaker presentations are available at https://www.nationalacademies.org/event/03-16-2020/realizing-opportunities-for-advanced-and-automated-workflows-in-scientific-research-second-meeting.

balance between access to and protection of data. In this report, we filter these issues through the lens of ARWs, recognizing that other National Academies committees have explored many of these issues in greater depth. Another topic that emerged at the March 2020 workshop and in recent literature is the role of scientific workflow engines as important enablers of effective development and implementation of ARWs. The committee recognizes this technology as critical for advancing the utilization of ARWs, and we refer to various tools and resources throughout this report. However, this more specific aspect of workflow management is distinct from our broader consideration of ARWs.

Our recommendations concern technical issues to be addressed through future research, as well as associated cultural, educational, and policy-related issues. The report is intended to create awareness, momentum, and synergies to realize the potential of ARWs in scholarly discovery. Issues and questions related to workflows for research that motivate our work included the following:

- How do ARWs affect the research process in various fields and disciplines?
- What are the barriers to implementing ARWs and how do they operate in different disciplines? These barriers include inadequate workflow literacy, resistance to adoption of scientific automation, insufficient appreciation of dangers of p-hacking, selection bias and model overfitting, and concerns about privacy protections of data and procedures (especially for cloud-based workflow systems).
- How does open research (encompassing open science and open scholarship), in the form of open availability of articles, data, code, and other research products, contribute to the utility and attractiveness of ARWs? And, conversely, how do ARWs contribute to the utility of open research?
- What technical and operational issues arise in implementing ARWs?
- What are the current regulatory enablers and barriers affecting the adoption of ARWs in various fields and disciplines?
- What considerations related to costs for equipment, software, staffing, and training come into play in the process of adopting ARWs?
- What are the implications of broader use of ARWs for educational approaches for students and faculty?
- Are there promising areas for investment and activity on the part of research funders and research institutions in the development and implementation of ARWs?
- And last, but critically, how do researchers need to evolve their practices to use ARWs in a manner that enables them to reap the benefits of automation while not losing the benefits of serendipity of discovery?

To address these questions, the report is organized as follows. Chapter 2 highlights the context for ARWs with a focus on the evolution and development of technologies and relevant policies that may facilitate or inhibit their greater use.

INTRODUCTION

Chapter 3 provides case studies of workflows across disciplines in the sciences and humanities; they are based on March 2020 workshop presentations, a review of the literature, and the committee's own experience. Chapter 4 looks at crosscutting issues across disciplines that ARWs can help with relative to research integrity, reproducibility and replicability, and dissemination. This examination feeds into Chapter 5's consideration of the barriers and opportunities across fields, to which the committee offers its recommendations. Chapter 6 presents the committee's findings and recommendations and offers concluding thoughts and potential next steps for researchers and institutions in both the public and private sectors, funders, and policy makers.

2

Context for Automated Research Workflows

THE OPPORTUNITY

As we enter the third decade of the 21st century, societal demands are converging that automated research workflows (ARWs) can help address. The volume and exponential growth of digital data and of the ability to mine and generate those data provide rich opportunities for progress. This growth has led to *quantitative* change in the way research is conducted. Pairing advances in artificial intelligence (AI), computing, and automation of laboratories and observations can also lead to a *qualitative* step change.

Technological developments go hand in hand with scientific progress, and advances in computing and automation are no exception. Computing plays a central role throughout research workflows, from computerized models used for simulation and prediction, to control of equipment and data analysis, to publication. Laboratories and observational devices are increasingly controlled by computers and are automated. For example, telescopes now are routinely controlled remotely by computers, and increasingly the observational process is automated following workflows predefined by humans. Biological and chemical laboratories are increasing use of microfluidic devices that enable automated experimentation at higher throughput and a faster pace than is possible by hand. Computers and automation have led to a quantitative increase in research productivity over the past half century.

As an indicator of growth, the "global datasphere"—the amount of data that is created, captured, copied, or consumed in a given year—was estimated to have reached 64 zettabytes (or 64 trillion gigabytes) in 2020 and is projected to grow to around 175 zettabytes by 2025 (Woodie, 2020). Computing performance has also exploded. The peak performance of the world's fastest computers has increased from 1 billion floating point operations per second (GFLOPS) in the 1980s to 10^8 GFLOPS in 2020.[1]

More radical changes are under way. Consider the evolution of the basic research workflow since the scientific revolution of the 17th century, which put science on an empirical footing. A scientist was not only to observe "nature in the raw," but also, in Francis Bacon's (1620) words, to "twist the lion's tail," that is, "manipulate our world in order to learn its secrets" (Hacking, 1983). Science since then has advanced through a virtuous circle in which measurements, observations, and, increasingly, simulations generate data. Harnessing the data leads to an update of existing models or the formulation of new models for the data (Figure 2-1). Knowledge generation can begin at any point in this loop, for example, either with a new model that prompts the generation of new data, or with new data that prompt the generation of a new model or the modification of an existing one.

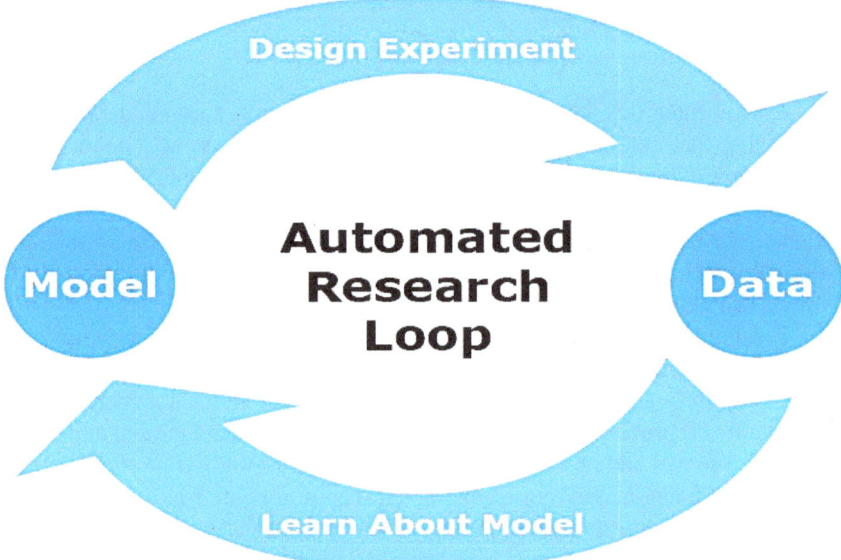

FIGURE 2-1 Knowledge discovery loop.
NOTE: Automated research workflows can automate and close the loop of scientific discovery. On one side of the loop, artificial intelligence (AI) and machine learning (ML) algorithms harness the experimental or observational data to learn about a model; on the other side of the loop, AI and ML are used to generate the study design for the next data collection. The loop goes on iteratively.

---

[1] For more information, see https://www.top500.org. Accessed April 19, 2022.

The process of starting from a model and devising an observational or experimental way of generating new data is called experimental design (with an experiment understood broadly to include, for example, the collection of observational data). As we use it here, experimental design does not need to imply constructing a comprehensive set of experiments to cover some experimental space or creating a new experimental procedure. It can be as simple as choosing a single set of experimental parameters (e.g., a particular combination of materials, chemical reactants, or drugs and targets) to test next using an established procedure. The process of using data to inform a model can be called learning about a model. We are using the term "model" broadly, to include instantiations of general theories (e.g., how the collision of black holes gives rise to gravitational waves according to the general theory of relativity) and empirical or semiempirical models (e.g., in economics or the environmental sciences). "Data" is similarly taken as an all-encompassing term—for example, data generated in simulation studies as well as in laboratory experiments, or in the digital humanities, original text sources, images, maps, social media, and much more.

The potential for greater progress lies in exploiting this proven success model of science, but accelerating it by orders of magnitude by iterating faster and continuously. Given a model, not only one but many (thousands and more) experiments may be designed automatically and can be optimized to be maximally informative about the model. The learning step can likewise be automated. Algorithms for designing experiments to be maximally informative and for learning about a model from data can become broad purpose and transcend individual disciplinary fields (in much the same way that least squares estimation has become broad purpose). ARWs can be structured so that data collection, analysis, and hypothesis revision and refinement are undertaken as a continuous process, with updates occurring as new data are generated or discovered (Gil et al., 2017).

However, models and data, and the means of acquiring data, will likely remain domain specific. That is, the edges of the graph in Figure 2-1 can be automated with methods that transcend individual disciplines, but the nodes—data and models—will remain domain specific. The loop remains open to human intervention, for example, to identify variables relevant to measurement and modeling, and to analyze serendipitous results.

To cast the discussion in modern machine learning (ML) terms, the closed-loop research workflow in Figure 2-1 encapsulates a form of reinforcement learning (Sutton and Barto, 2018), in which a model is used to design a manipulation or observation of an environment to generate data (experimental design), and the model subsequently learns from the data so generated. Reinforcement learning in essence is how science has progressed for centuries. Similar gains from rapid iteration in what may be understood to be a reinforcement learning loop—but may include a variety of techniques from ML, Bayesian learning, and experimental design—are now possible in some scientific fields.

A new generation of workflows is making extensive use of AI, ML, and, in general, automation. (See Box 2-1 for working definitions for these and other

BOX 2-1
Key Terms Used in This Report

Automated Research Workflows
ARWs integrate computation, laboratory automation, and tools from artificial intelligence in the performance of tasks that make up the research process, such as designing experiments, observations, and simulations; collecting and analyzing data; and learning from the results to inform further experiments, observations, and simulations.

Artificial Intelligence
While specific definitions vary, artificial intelligence is, generally speaking, any method for programming computers to enable them to carry out tasks or behaviors that would require intelligence if performed by humans (NAS, 2018).

Cyberinfrastructure
The concept of cyberinfrastructure first emerged in the late 1990s and early 2000s. The term has come to encompass a spectrum of computational, data, software, networking, and security resources, tools and services, and computational and data skills and expertise that can be seamlessly integrated and used, and collectively enable new, transformative discoveries across science and engineering (NSF, 2019).

Machine Learning
Machine learning draws from a variety of fields, including computer science, statistics, engineering, cognitive science, and neuroscience. Researchers in machine learning develop both the mathematical foundations and the practical applications of systems that learn from data (NAS, 2018). In the context of this report, we use the term *machine learning* broadly, to comprise any form of learning from data, be that Bayesian learning about parameters, parametric functions, or nonparametric functions in scientific models or learning with artificial neural networks.

Open Research
Open Research (which incorporates Open Science and Open Scholarship) aims at increasing research quality, boosting collaboration, speeding up the research process, making the assessment of research more transparent, promoting public access to scientific results, as well as introducing more people to academic research. It is a set of principles and practices that fosters openness throughout the entire research life cycle (EC, 2018; NASEM, 2018).

Reproducibility and Replicability
Reproducibility is obtaining consistent results using the same input data; computational steps, methods, and code; and conditions of analysis.
Replicability is obtaining consistent results across studies aimed at answering the same scientific question, each of which has obtained its own data. Two studies may be considered to have replicated if they obtain consistent results given the level of uncertainty inherent in the system under study (NASEM, 2019).

> **BOX 2-1 Continued**
>
> **References**
>
> EC (European Commission). 2018. OSPP-REC, EC open science policy platform recommendations 2018. Available at https://ec.europa.eu/info/research-and-innovation_en#view=fit&pagemode=none. Accessed May 21, 2021.
> NAS (National Academy of Sciences). 2018. *The frontiers of machine learning: 2017 Raymond and Beverly Sackler U.S.-U.K. Scientific Forum.* Washington, DC: The National Academies Press. https://doi.org/10.17226/25021.
> NASEM (National Academies of Sciences, Engineering, and Medicine). 2018. *Open science by design: Realizing a vision for 21st century research.* Washington, DC: The National Academies Press. doi: 10.17226/25116.
> NASEM. 2019. *Reproducibility and replicability in science.* Washington, DC: The National Academies Press. doi: 10.17226/25303.
> NSF (National Science Foundation). 2019. *Transforming science through cyberinfrastructure.* Available at https://www.nsf.gov/cise/oac/vision/blueprint-2019/nsf-aci-blueprint-v10-508.pdf. Accessed July 12, 2021.

key terms used in this report.) AI and ML are increasingly used as components of ARWs across many domains: examples include understanding protein folding in biology and analyzing sparse data in the geosciences (Gil et al., 2019: Hey et al., 2020). Beyond that, AI and ML are beginning to be used to automate the design and operation of elements that are traditionally considered part of the workflow itself, such as the design of experiments (Deelman et al., 2019). This offers the opportunity to produce a next generation of workflows that are dynamic, intelligent, and self-governing. Distinct uses of AI and ML and workflows play a role in the many phases of a research project (e.g., planning, exploration, scale-up, and publishing). AI and ML techniques deployed within ARWs not only can drive an experiment and mine the literature to suggest future experiments, but also may enhance research reliability and productivity by facilitating the reuse of workflows and improving the ability of researchers to monitor workflow execution and detect anomalies (Deelman et al., 2019). Similar concepts in nonscientific domains are being called *intelligent* or *cognitive* workflows (Bellissimo, 2019).

ARW opportunities are vast but are accompanied by technical and mathematical challenges and, perhaps even more so, by organizational, economic, policy and political, social, and incentive issues. Wide adoption of ARWs requires consideration of these interrelated concerns. In the United States, the leadership for advancing computing for research and higher education has come from the National Science Foundation, the Department of Defense (DOD), the Department of Energy (DOE), and the National Institutes of Health (NIH), sometimes coordinated by the Office of Science and Technology Policy (OSTP). Evolution of the technology alone is not enough to ensure adoption, access, and meaningful transformative uptake by the various specialized science communities. What is required, and what these agencies have nurtured in the past, is the building of compelling visions of possibilities and scientific impact and then using these visions, backed by targeted funding, to motivate innovation in the science communities they serve. Federal financial support can stimulate related support from the private sector, nongovernmental foundations, and universities. Examples include cyberinfrastructure research and development (R&D), pilot application projects, and the related human resource development. The discussion of use cases, barriers, and opportunities in this report aim to illustrate how ARWs will become essential to the exploration of the frontiers of discovery and the grand challenges facing our world.

BUILDING AUTOMATED RESEARCH WORKFLOWS: CURRENT STATE OF THE ART

As outlined above, the confluence of several technological advancements is driving the development and implementation of ARWs. Fully realized ARWs are not common at present, and so this study examines how and where progress is being made in areas such as advanced computation, use of workflow management systems and notebooks, laboratory automation, and use of AI as a workflow component as well as in directing the "outer loop" of the research process. These separate developments are producing positive changes in research processes. For example, the broader use of workflow management systems is delivering gains in reproducibility and reliability (Ferreira da Silva et al., 2021a,b). Likewise, the use of AI and ML in a range of disciplines is proceeding rapidly, whether they are used in conjunction with workflow management systems (Royal Society and Alan Turing Institute, 2019).

Scientific Workflow Engines and Related Software Tools

The concept of a "workflow" as a series of computational steps in the scientific discovery pipeline is not a new one. In a sense, any computational analysis pipeline that involves multiple stages and dependencies across those stages can informally be considered a workflow. These steps are often linked and chained together by a set of scripts to provide some degree of automation.

As the nature of research problems and the cyberinfrastructure platform for exploring them have become more powerful and complex, scientific workflow engines have played a crucial role in harnessing and coordinating distributed computing and data resources. Scientific workflow engines are software tools that capture the computational analysis pipeline of a research project, providing provenance tracking and other functions that facilitate automation, reproducibility, and reusability. They are proliferating; one community effort has compiled a list of more than 280 scientific workflow engines and acknowledges that its list is probably incomplete (GitHub, 2021). Table 2-1 provides a representative list.

TABLE 2-1 Examples of Workflow Engines and Related Tools

Airflow	https://airflow.apache.org
Bigtable	https://cloud.google.com/bigtable
Chimera	https://github.com/hysds/chimera
Cromwell	http://cromwell.readthedocs.io/
Cyverse Discovery Environment	https://cyverse.org/discovery-environment
Fireworks	https://materialsproject.github.io/fireworks
Hadoop	https://hadoop.apache.org
Galaxy	https://galaxyproject.org
iRODS	https://irods.org
Jupyter	https://jupyter.org
Kepler	https://kepler-project.org
Nextflow	https://www.nextflow.io
Open Science Framework	https://osf.io
Luigi	https://luigi.readthedocs.io/en/stable/workflows.html
Parsl	http://parsl-project.org
Pegasus	https://pegasus.isi.edu
Snakemake	https://snakemake.readthedocs.io/en/stable
Spark	https://spark.apache.org
Starfish Storage	https://starfishstorage.com
Wolfram	https://www.wolframcloud.com

NOTE: Many of these tools are tracked by workflow community initiatives such as WorkflowHub (https://workflowhub.eu) and WorkflowsRI (https://workflowsri.org).

The recent report from the Workflows Community Summit held in January 2021 emphasizes the need for building a workflows community (Ferreira da Silva et al., 2021a,b).

Scientific workflow engines formalize a workflow construct, in which a user defines a set of steps and the dependencies between those steps through configuration files and code. They are key to enabling automation and reusability in the scientific pipeline. This allows for more complex relationships such as directed acyclic graphs or loops to be expressed in the workflow. The same workflow can then be reexecuted under different initial conditions, against multiple data sets, and at different scales. These workflow engines have become key in large-scale analysis pipelines, both in academia (e.g., Kepler, Parsl, Pegasus, Fireworks, and Cromwell) and the commercial sector (e.g., Airflow and Luigi). Some of these tools are used to scale up AI or ML runs for automated parameter optimization or training models on very large data sets. There are also several distributed computing and automation frameworks with a narrower focus that capture specific execution patterns. These may also be intrinsically part of the "workflow." Examples include tools such as Spark or Hadoop that enable a large number of data processing tasks at scale, or cloud data stores such as BigTable that can execute queries across a large distributed data set.

Also worth mentioning are tools that implicitly encapsulate interactive workflows—"interactive notebooks" such as Jupyter, Wolfram, MATLAB, and

RStudio provide a user-friendly interface to capture a set of discrete analysis steps in a single document, where each step can be run interactively, and intermediate results between steps can be examined and visualized. The popularity of these approaches is a testament to the fact that workflows are quite ubiquitous in scientific computing even if scientists do not formally use the term. These digital notebooks may serve as an on-ramp for students and researchers to a new generation of production-level, customizable tools that would be widely adopted.

Looking ahead, there is an opportunity for workflow engines to integrate with the "outer loop" of research discovery, where some level of input or feedback is required to drive the process, from human input, an external interaction with the physical world, or a computational process outside the automation engine. Electronic notebooks such as Jupyter enable human-in-the-loop interactivity and have thus provided scientists with the ability to tie human insight and iteration into this process. AI and ML workflows in particular have benefited from this, because many of the steps (model training, hyperparameter optimization, etc.) are human-centric processes that involve examination of intermediate results to inform future iterations. Building on the interactive workflow paradigm, Galaxy, with over 30,000 users as of April 2021, is an example of a workflow system where user setups are recorded, and where notebooks and workflows may be routinely intertwined (Galaxy, 2021a).

We also note the emerging "Canonical Workflow Frameworks for Research" initiative, to capture canonical workflows and workflow patterns that are built on some of these technologies. This effort seeks to improve reuse of workflow components, increase efficiency of workflows, and allow incorporation of machinery that automatically generates findable, accessible, interoperable, and reusable (FAIR) data (Hardisty and Wittenburg, 2020).

As the next generation of scientific workflow engines expands, automation of the scientific process can lead to a step change in the rate of discovery in many fields. Room for serendipity and human ingenuity remains essential, and so interactivity and integration of external input must be a core part of the system. In essence, we need to go beyond a closed automation loop and enable interaction points for modifying and driving the system and identifying relevant variables. In other words, next-generation workflows require tools that provide both interactivity and automation at scale.

Data Resources

In 2014, academic and private researchers interested in overcoming obstacles for data discovery and reuse met in the Netherlands. From that meeting grew a set of principles calling for all research objects to be FAIR. GO FAIR[2] formed as a stakeholder initiative to implement these principles and is funded by the Ministries

---

[2] See https://www.go-fair.org/. Accessed August 21, 2020.

of Science in France, Germany, and the Netherlands. The European Open Science Cloud (EOSC), based on the FAIR principles, has, as one of its priority actions, to "develop and sustain core data assets for the EOSC and make them available to the community under well-defined conditions. These assets may include workflows, analytics, programmes and notably existing data sets with FAIR status" (Wilkinson et al., 2016). While the FAIR principles call for all research objects to be FAIR, they are somewhat specific to data, and there have been subsequent efforts to address other types of objects, such as research software, FAIR workflows, and for the start of an effort focused on ML models (Goble et al., 2020; RDA, 2021a,b).

A key aspect of FAIR data that enables ARWs is machine readability. Making more FAIR data available allows ARWs to find the data that are relevant to a research task in question and incorporate these data into the analysis. Wider reuse encourages researchers to make more FAIR data available, creating a virtuous cycle.

Workflows can adhere to and advance FAIR data principles "by processing data according to established metadata, creating metadata themselves during the processing of data, and by tracking and recording data provenance" (Goble et al., 2020). Properly designed ARWs support FAIR data principles since they can capture the associated metadata and provenance necessary to describe their data products in a formalized and completely traceable way. They can provide more accurate curation of the data to support both data reuse and data review (to support assessment of reproducibility or robustness and of conclusions), significantly reducing the time and hence cost of making data FAIR. Creating workflows can be research products in their own right, encapsulating methodological know-how that needs to be published, accessed and cited, exchanged and combined with others, and reused as well as adapted. Data resources and needs related to specific domains are discussed in Chapter 3.

Progress is being made in the number and diversity of domain-specific and general data repositories that support FAIR principles and provide archival functionality for long-term access to data and related research objects. Examples can be found in the Registry of Research Data Repositories.[3]

Progress in Domain-Relevant Artificial Intelligence and Machine Learning

Another key factor in building ARWs is the continued advances in learning algorithms for specific domains. New and better algorithms, defined as "encoded procedures for solving a problem by transforming input data into a desired output, based on specified calculations and procedures," have been fundamental to the progress of ML across many domains within and outside of research (Gillespie et al., 2014; NAS, 2018). A report from a 2018 DOE workshop on basic research needs for advancing ML in science suggests a focus on "creating domain-aware,

---

[3] See https://www.re3data.org.

interpretable, and robust ML formulations, methods, and algorithms" (DOE, 2019). It identifies incorporating domain knowledge into ML as a key challenge:

> In many of the most successful ML examples, such as image recognition, system developers know the "ground truth" sufficiently well to check the results, often even while training the models. Almost by definition, the most interesting scientific applications of [scientific ML] are those, such as materials discovery or high-energy physics, where the answers are unknown beforehand or the results of an automated system are not easily verified.

Understanding and managing the interplay between models derived from domain knowledge, ML, and how the system iteratively drives experimental design constitute a continuing task for ARW development across domains.

IMPLEMENTING AUTOMATED RESEARCH WORKFLOWS: A CHANGING SCIENTIFIC PARADIGM

Over the past two decades, scientific workflow systems have matured as powerful tools, especially for "resource allocation, task scheduling, performance optimization, and static coordination of tasks on a potentially heterogeneous set of resources" (Altintas et al., 2019). As a platform for these software capabilities, existing cyberinfrastructure provides important components that can be incorporated into ARWs to translate new advances into repeatable, scalable solutions (Altintas et al., 2019).

Much of the progress in developing workflow management systems was carried out in the mid-2000s (Taylor et al., 2007). Their use has been limited initially by factors such as difficulties in incorporating human decision making into the loop, difficulties in adding new components, lack of interoperability, and the need for large commitments of time and effort to manage and maintain them. The developments that are combining to encourage greater use include the maturation of the systems, AI as a component in the workflow, better interoperability of systems and components, and the promise of open science and FAIR principles to raise the value of workflows broadly and evolve into ARWs.

Scientific workflow engines have historically targeted applications in scalable computing where users chain together multiple steps in a complex computational process (e.g., job submission to a supercomputer, access to a database, execution of a web service) to express a dataflow. Such use of workflow engines is akin to use of an existing recipe, where the workflow designer, often an individual using a graphical or script-based user interface, programs a known dataflow of tasks for scale and reuse. However, solving the biggest problems of our time requires two main changes from this use of workflows: (1) shifting from an individual user to teams for designing the workflow and (2) capturing the evolution of a workflow and the data-driven explorations to make them later scalable in various forms. Reproducibility of the process under this new paradigm is an additional consideration.

From Workflow to Teamflow: Emergence of Teams as the User

Solving big, complex problems requires end-to-end workflows for integrated management of many technical steps in addition to extensive knowledge of the application domain. These integrated steps require expertise from multidisciplinary team members to collaborate on (1) methods to manage, integrate, and interpret "big" data; (2) modeling and simulation tools executing on scalable computing platforms; (3) methods and interfaces for domain-specific analysis, communication, and visualization of results; and (4) technologies to make the process FAIR, as well as portable, transparent, repeatable, and reproducible (GO FAIR). Such multidisciplinary collaboration shifts the paradigm from individual to team science, that is "research conducted by more than one individual in an interdependent fashion, including research conducted by small teams and larger groups" (NASEM, 2015). Team science requires tools for managing, capturing, and advancing team collaboration, contribution, and communication as an open process, in addition to the discovery process and its reproducibility.

As an example, the scenario in Figure 2-2 illustrates the scientific collaboration process among team members with complementary scientific and technical expertise in areas ranging from scientific domain expertise to data engineering, data analysis, and computational science. The research ideation and design is typically initiated by a domain scientist, triggering modeling and simulation. The experimentation toward scalable modeling and simulation is often complemented by the work of a data engineer to ensure that data from simulations and experiments are "acquired, modeled, and queried effectively for analysis and computational modeling. The data scientist generates insights from the data so that the computational model can be parameterized effectively" (Altintas et al., 2019). For example, the data and computational scientists might collaborate on parameter estimation, ML, or data assimilation methods so that the computational model benefits from the data analysis. The team members develop a process through exploratory activities and iterative communication, and scale the process to execute in an automated fashion on advanced computing environments once the exploratory activity results in a mature workflow.

These roles may have some overlap across individuals, and there may be other work functions to consider. In addition, these roles are evolving and there are differences between disciplines in how they are implemented. For example, the role of data curator is not currently established in all disciplines. There may be hybrid roles such as "workflow system administration" that could span the tasks of a data engineer and software developer. The broader point is that multiple people are involved in the scientific discovery process across the workflow.

To be sure, transitioning from individual to team-oriented research not in itself may be a central obstacle to building the human collaboration needed to implement ARWs. For example, each of the research contributors shown in Figure 2-2 may come to the collaboration with a set of tools with which they

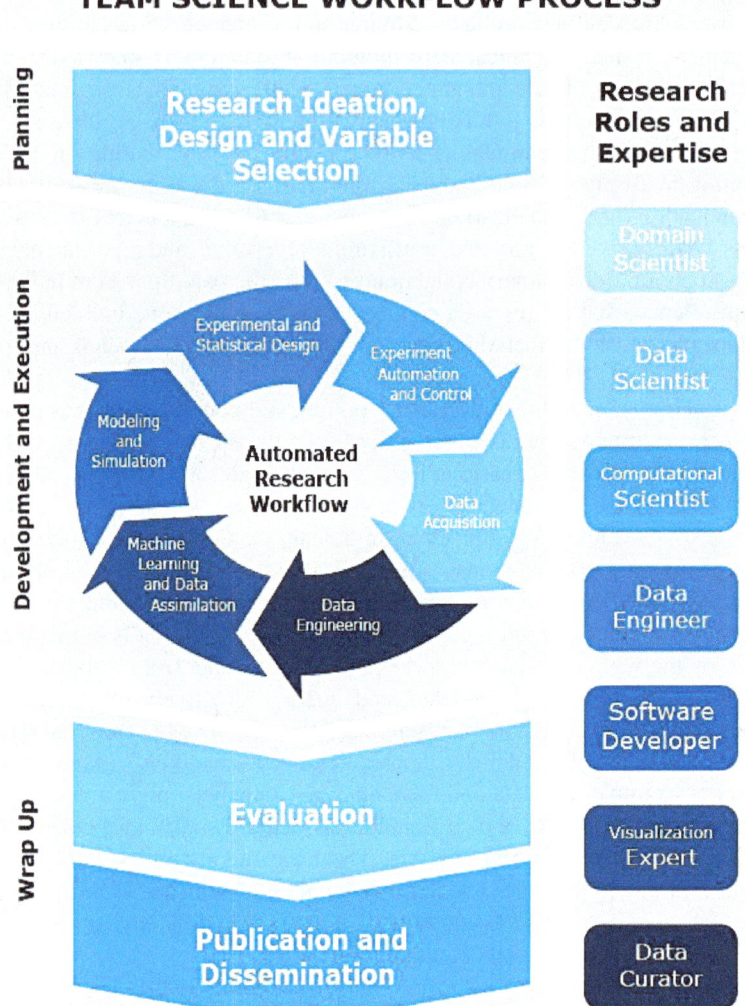

FIGURE 2-2 Team science workflow process.

are familiar or that work with the equipment they have at their disposal. Most workflow systems require that the collaboration adopt a specific set of tools and specific methodology for its research. That is, the workflow engines or other enabling tools may embody ways of conducting the work that need to be aligned with the human participants. Successful groups will need to agree on a common approach, such as building up an environment around the tools they already use.

From Exploratory Activity to Scale

Once a research team agrees on its research methods through exploration, there is often a need to scale up execution processes with more data or for larger parameter sets requiring automation and control. A big challenge in building ARWs is to sustain the linkage between the exploratory activities and the automated scalable process. It is inherently more difficult to fund development and maintenance of production-quality software (workflow engines, automated tools, etc.) that can be used broadly than to develop new software as part of a research project that may not be used outside that project.

Often, following the exploratory activities, the process is reimplemented for scale as an automated process instead of reusing the software developed during the exploration phase. Because the exploratory and scalable components are separate, iterations between exploratory and scalable automated activities become difficult to manage (Reiter et al., 2021). ARWs can strengthen the link between exploration and scale in three ways: (1) capture key information (e.g., performance, accuracy of individual steps) during the exploration to enable seamless scalability of the final process; (2) enable auto-scalable converged application through communications with data and computing middleware; and (3) optimize resources and dynamically adapt to the changes in the underlying cyberinfrastructure (Altintas et al., 2019). Automating data collection in order to analyze and use the data is key to building effective systems that bridge exploratory and scalable activities, make the workflows more useful and align with the way teams of researchers collaborate, and develop integrated applications.

Reproducibility of the Process and Team Science

Scientific workflow engines potentially provide "a programming model for deployment of computational and data science applications on all scales of computing and provide a platform for system integration of data, modeling tools, and computing while making the applications reusable and reproducible" (Altintas, 2018). Many research workflow systems today provide capabilities for provenance tracking, repeatability, and partial reproducibility support. However, the shift from individual workflow development to team science also creates the need for workflow systems to capture the process for validation, seamless integration, and repeatability of the team's activity.

Figure 2-3 illustrates in lighter blue the system hierarchy supporting the discovery loop by which the research team interacts with the scientific workflow engine and other software tools to run ML or AI algorithms or methods in a computing infrastructure using data to learn about the model and then to design new experiments based on what is learned. Increasingly these research teams are distributed across disciplines, organizations, and geography. The dark blue lower box lists the best and responsible practices in research that should apply across all levels, as data and code are used throughout.

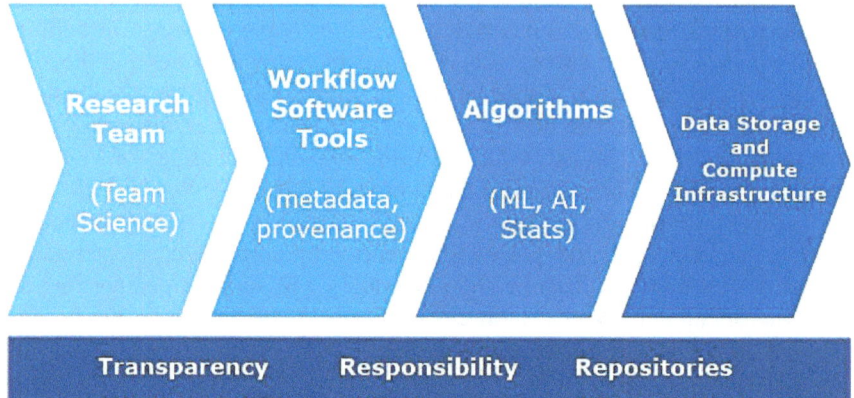

FIGURE 2-3 ARW components and context.
NOTE: The research team interacts with the workflow tools to run ML or AI algorithms and methods in a computational infrastructure to learn about the model using data and then to design new experiments based on what is learned. Responsible and best practices in scientific discovery apply across all components, as data and code are used throughout. All these components should be funded and sustained for the automated closed loop to work and advance.

POLICY AND INDUSTRY CONTEXT FOR AUTOMATED RESEARCH WORKFLOWS

Public Policy Readiness

Policy makers and funding agencies in the United States and Europe have articulated a research vision at a scale and complexity that implies robust support for the development and sustainability of ARWs. That is, while not explicitly singling out "support for ARWs," they point to the societal and economic benefits that AI and ML can bring about. In most cases, achieving these benefits requires making use of ARWs as described in this report, and funding has begun to address this reality.

At the committee's March 2020 public workshop, Kelvin Droegemeier, then director of the White House OSTP, stressed the importance of AI as one of four "Industries of the Future," together with advanced manufacturing, quantum information science, and 5G network capability. The 2021 federal budget included $868 million (a 76 percent increase) to the National Science Foundation for AI-related grants and interdisciplinary research initiatives; $125 million to DOE's Office of Science on AI research; $50 million to NIH for research on chronic disease using AI and related approaches; $459 million for AI R&D at the Defense Advanced Research Projects Agency (DARPA); and $290 million for DOD's Joint AI Center (CRS, 2020).

Experts from several agencies provided context to the committee about how they envision a role for workflows. For example, at DOE, the Artificial Intelligence & Technology Office has been established to "transform DOE into a world-leading AI enterprise by accelerating the research, development, and adoption of AI," with ARWs as a core part of its efforts (Kusnezov, 2020). Within DOD, DARPA funds projects that rely on or support the development of workflows, including those related to automating molecular discovery and an AI exploration program (Russell, 2020). In addition, the National Science Foundation's Big Ideas program encompasses 10 areas in which it is investing in pioneering research, addressing big ideas, and pilot activities. While workflows are fundamental to many of these 10 areas, 2 are particularly salient here:

- *Growing convergence research* focuses on grand challenges that require multidisciplinary approaches: "From its inception, the convergence paradigm intentionally brings together intellectually-diverse researchers to develop effective ways of communicating across disciplines by adopting common frameworks, and a new scientific language."
- *Harnessing the data revolution* involves support for a "cohesive, federated, national-scale approach to research data infrastructure, and the development of a 21st-century data-capable workforce."

The priority here and across the government is on identifying large, complex problems and then finding the workflows and other processes to solve them, rather than first developing technologies and then finding an application.

On January 1, 2021, Congress passed the National Artificial Intelligence Initiative Act (NAIIA) as part of the National Defense Authorization Act.[4] The act's language reflects many of the issues raised by committee members and presenters. It sets out a major role by the federal government to incentivize and support AI, including access to data sets, computing resources, and real-world test environments; improved standards and benchmarking for AI systems; and removal of barriers to interdisciplinary collaboration. It also recognizes the need to educate an AI-savvy scientific workforce. In June 2021, OSTP announced the formation of the National Artificial Intelligence Research Resource Task Force as part of implementing the NAIIA.

> The Task Force will provide recommendations for establishing and sustaining the NAIRR, including technical capabilities, governance, administration, and assessment, as well as requirements for security, privacy, civil rights, and civil liberties. The Task Force will submit two reports to Congress that together will present a comprehensive strategy and implementation plan — an interim report in May 2022 and final report in November 2022. (OSTP, 2021)

---

[4] 15 U.S.C. Chapter 119, https://uscode.house.gov/view.xhtml?path=/prelim@title15/chapter119&edition=prelim.

The European Union has prioritized open research data in its research and innovation policy making, driven by the need for reproducible science, appreciation of data as a strategic asset (akin to how a country might value its oil reserves), and a desire to avoid overdependence on companies such as Google and Facebook (Burgelman, 2020). In 2019, the EU's Directorate-General for Research and Innovation conducted a cost-benefit analysis that estimated that *not* having FAIR data would cost Europe about $10 billion per year (EC, 2019). This has led to development of the EOSC as a shared infrastructure to provide access to data repositories and resources such as cloud services, high-performance computing, and data analysis tools (EOSC, 2020).

To back up policy priorities, the recently concluded European Union's research and innovation programme, Horizon 2020, provided €80 billion in funding between 2014 and 2020, including emerging technologies, e-infrastructure, and advanced computing. It increased its investments in AI by 70 percent from 2018 to 2020, to about €1.5 billion, aiming to increase total investment in AI (public and private combined) to €20 billion per year by the end of 2020 (EC, 2021a). In addition, the EU's Artificial Intelligence and Blockchain Investment Fund is set up to make €100 million available to companies, with the idea that these funds will leverage additional private- and public-sector support. Moving forward, the new EU Framework programme, Horizon Europe, will provide €95.5 billion in research funding for 2021–2027. As part of this, the European Commission announced a new €7.5 billion Digital Europe Programme[5] with focus areas in supercomputing, AI, cybersecurity, advanced digital skills, and wider use of digital technologies.

The United Kingdom Industrial Strategy aims for the UK to become the "world's most innovative economy," with R&D investment to reach 2.4 percent of gross domestic product by 2027. Realizing this vision includes developing the infrastructure to support it—ranging from physical facilities such as research ships and satellites, to archives and repositories, to cyberinfrastructure. In 2020, United Kingdom Research and Innovation (UKRI), the major public funder of research, published a review of the country's research and innovation infrastructure to guide funding decisions and other priorities until 2030 (UKRI, 2020). In addition to computational and e-infrastructure as one of the review's six themes, the potential contributions of intelligent workflows are suffused throughout the other themes (biological sciences, health, and food; physical sciences and engineering; social sciences, arts, and the humanities; environment; and energy). It calls for a "sustained multi-year investment" in supercomputing; data infrastructure; cloud computing; network and cybersecurity; authentication, authorization, and accounting infrastructure; and software and skills. In January 2020, the government allocated £300 million to UKRI to fund research infrastructure. Many UK research institutes and infrastructures are also playing key positions and providing pivotal input into the EOSC and have led the initial computational development

---

[5] See https://ec.europa.eu/digital-single-market/en/europe-investing-digital-digital-europe-programme.

work via the Science and Technology Facilities Council, a part of UKRI (UKRI, 2017). The UK is also a world leader in open research data, for example, with the UK Data Archive that has retained social science and humanities data for almost 50 years, and the delivery of a Concordat on Open Research Data (UKRI, 2016).

These public affirmations of support and new banner initiatives related to AI, infrastructure, data sharing, and related capabilities in the United States, Europe, and the UK do not, in and of themselves, guarantee adequate funding. However, such efforts provide a positive policy climate in which advancements may flourish. Note that China is also making investments in cyberinfrastructure. For example, the CSTCloud (China Science and Technology Cloud) is emerging as a national infrastructure of data and computation for accelerating science discovery.

Industrial Use of Workflows

This discussion of industrial use of workflows focuses primarily on research applications. Industrial development and use of computational workflows extends beyond—and predates—the use of workflows in the realm of scientific research. Computational workflows in industry have automated many business processes such as "generating an email response when a customer fills out a request form, transaction processing or communicating with multiple databases while processing an insurance claim" (IBM, 2021). There are barriers to translating practices and tools developed for computational workflows used to automate business processes to research applications. Business workflows have been built to perform a well-characterized series of tasks accurately and repetitively, while research applications may need to accommodate the collection and analysis of various types of experimental data (Barga and Gannon, 2007). Still, as ARWs become more widely used, it should be expected that more characteristics of industrial workflows—security and integrity in particular—will be valued and adapted at a larger scale into ARWs.

Several participants at the March 2020 public workshop commented that industry has been more open than academia to the use of ARWs in research, perhaps because workflows are common on the business side of their operations or because of bottom-line imperatives to maximize efficiencies. Companies are involved with workflows in two primary ways: using them in their own research (e.g., pharma companies doing high-throughput screening) and developing and marketing workflow tools, services, and training. These might be offered as part of a suite of products (e.g., IBM, Amazon Web Services), while a few have made scientific process development their primary focus. Several companies have also developed products to automate and accelerate the publishing process, as highlighted in Chapter 4. Examples of academic uptake of commercial workflows include KNIME, RapidMiner, and Rabix.

Riffyn was launched in 2014 with the mission to integrate process data from scientific experiments for ML. According to CEO Timothy Gardner, Riffyn was intentionally set up as a for-profit company to capitalize on "resources, sustainability, and usability," which he said could best be achieved through a for-profit structure. One example related by Gardner is a biotech firm that used Riffyn's intelligent workflows and other processes to accelerate its development of advanced yeast strains. The company brought four strains to market in 18 months, which had both discovery-enhancing and financial benefits. Although Riffyn provides its product at no cost to academic researchers, Gardner noted that uptake is low, which he attributed to the educational and cultural barriers discussed in Chapter 5.

Industry has some priorities that are distinct from those of governments or academia. For example, industry often uses open academic research data, but opens its own research data much less often. Companies may also use proprietary workflow tools that store and manage data in nonstandard proprietary formats. Since there is little incentive for toolmakers to agree to standards among themselves, researchers may be unable to access or utilize data even if they are technically open. Several presenters at the workshop have worked in academia and the private sector and provided ideas on how to strengthen links across different organizational cultures and constraints.

3

Automated Research Workflows in Action

The committee explored a number of examples of research drawn from a range of disciplines in order to understand how automated research workflows (ARWs) are being implemented as well as the factors that facilitate and hinder their adoption. Most of the use cases were discussed at the March 2020 workshop, while several others were examined through a literature review and interviews. The use cases were selected to provide a broad perspective of ARW implementation. In some cases, the implementation of ARWs is fairly advanced, whereas in others only certain components or aspects of ARWs—such as advanced computation, use of workflow management systems and notebooks, laboratory automation, and use of AI as a workflow component as well as in directing the "outer loop" of the research process—are currently being utilized.

The purpose of examining these specific areas of research was not to develop a comprehensive census of projects and initiatives, and these examples do not represent a complete picture of relevant work in general or in the disciplines that are represented. Rather, the goal was to learn through these use cases how ARWs are changing research at present and to understand their future potential. This exploration of use cases shows the tremendous potential of next-generation workflows to enable and catalyze new approaches to research across multiple, varied fields. Table 3-1 provides a snapshot of the issues and barriers raised in the use cases. Several challenges identified across the domains, such as institutional culture and training needs, are discussed further in Chapter 5.

TABLE 3-1 Use Cases Illustrating the Promise of ARWs and Challenges to Implementation

Use Case	General Characterization	Opportunities	Challenges and Barriers
Astronomy	History of big data, open data, and cyberinfrastructure is creating the potential for ARWs	Opportunities to use machine learning are expanding	• Tradeoff exists between collection of new data vs. stewardship of existing data • New data infrastructure and services are needed to provide access to FAIR data
Particle physics	Established history of using big data and creating sophisticated cyberinfrastructure exists	New opportunities exist for artificial intelligence (AI) approaches, such as simulation-based inference	• Planning and construction of large facilities to enable broader use of data outputs needs to be rethought • Reuse of workflows needs to be encouraged • Incentives and culture around data sharing vary by subfield
Materials science	Approaches are appearing that integrate robotic laboratory instruments, rapid characterization, and AI	Opportunities exist to link workflows and implement closed-loop systems	• Ability of humans to make unexpected observations needs to be preserved • Data sharing and access are inadequate • Shortage of researchers who can bridge the gap between experimentation and software development exists • Lack of a supportive culture within the community translates into weak incentives
Biology	Biomedical research has been overturning reductionist paradigms as they have lost predictive power, paving the way for empirical data to guide discovery	Potential for drug discovery approaches using automated experiments and AI is growing	• Shortage of experts who can bridge the gap between disciplinary and lab automation expertise exists • Capability for real-time data sharing is needed
Biochemistry	Tighter coupling between data science and chemical synthesis can accelerate optimization in drug discovery	Automation of high-throughput synthesis and screening can accelerate the design–make–test cycle	• Nonproprietary data are insufficiently characterized, so labs making advances in this area have to generate their own • Cultural barriers exist in the field (e.g., chemical synthesis has been an artisan process)

TABLE 3-1 Continued

Use Case	General Characterization	Opportunities	Challenges and Barriers
Epidemiology	COVID-19 experience has catalyzed initiatives to implement workflows	New data resources could be used to better understand disease interactions and improve treatments	• Data that are not generated through a randomized controlled trial might be biased • Confirming the quality and provenance of clinical data is difficult
Climate science	Improving climate models partly depends on the ability to understand and simulate small-scale processes	ARWs hold the promise of rapidly improving the accuracy of model predictions	• Climate sciences and weather data have traditionally been open, but wider use of commercial data obtained through restrictive licenses is a threat
Wildfire detection	Interdisciplinary field combines modeling, remote sensing, data science, and other fields	New tools and capabilities using AI can support decision making by public- and private-sector users	• Development of an integrated environment is needed to pull data from various resource monitoring tools and convert them into predictive intelligence
Digital humanities	Increasing use of computational tools and large data sets is expanding the sorts of research questions that can be addressed	New digital data and AI tools can be used to analyze big data (e.g., Latin texts)	• Human in the loop is needed to provide training material for AI • Gap exists between traditional research and new tools needed by next-generation scholars
Social and behavioral sciences	Access to large amounts of high-quality data is available at relatively low cost, transforming a number of fields	Real-time access to data and analysis can deliver actionable information	• Growing size and dispersed nature of data sets and need to ensure integrity of personal information pose challenges • Improving metadata is necessary for robust findings and reuse of data

PHYSICAL SCIENCES

Astronomy

Astronomy has long been driven by big data, and the amount of data available from telescopes on the ground and in space is rapidly increasing. Automation and increasingly fine controls on telescopes now make it possible not only to collect data after a careful (human) target selection, but also to close the loop between data acquisition and selecting the next target that is optimally informative given the observational constraints and scientific objectives. Astronomical surveys, such as the Large Synoptic Survey Telescope, Palomar Transient Factory, Catalina Real-Time Transient Survey, and Zwicky Transient Facility, have demonstrated the effectiveness of machine learning (ML) for extracting knowledge from astronomical data sets and streams (Juric et al., 2019). ML has also been used for new astronomical discoveries, including "finding new pulsars from existing data sets; identifying the properties of stars and supernovae; and correctly classifying galaxies" (Royal Society and Alan Turing Institute, 2019). The University of California, Santa Cruz, recently developed a new computer program called Morpheus that "can analyze astronomical image data pixel by pixel to identify and classify all of the galaxies and stars in large data sets from astronomy surveys" (Stephens, 2020).

At the workshop, Szalay (2020) discussed how automated workflows in astronomical research have evolved over the last 20 years and might advance in the future. The Sloan Digital Sky Survey (SDSS),[1] one of the largest and most often cited surveys in the field, has revolutionized the interactions between a telescope, its data, and its user communities (NAS, NAE, and IOM, 2009; NASEM, 2018a; Szalay, 2017). All SDSS data, including SDSS-I (2000–2005), SDSS-II (2005–2008), SDSS-III (2008–2014), and SDSS IV (2014–2020), are publicly available, and SDSS-V will start observations in summer 2020. There are more than 8,000 articles published with well over 400,000 citations, 3.5 billion web hits during the past 18 years, 480 million external SQL queries, and 7 million distinct users with 15,000 astronomers (Szalay, 2020). The data obtained from the project were available at the SDSS SkyServer (a public database managed by the Johns Hopkins University), which has been converted to the SciServer framework, a more collaborative platform with system-capturing interactivity and increasingly complex analysis patterns (Szalay, 2020). SciServer currently supports additional scientific disciplines, such as genomics, turbulence/physics, material sciences, and social sciences.

Szalay (2020) stressed the importance of prioritizing data, which requires tradeoffs between scientific value and costs of data management or preservation. There is a need for more relevant data (instead of just larger amounts of data) and

---

[1] See https://www.sdss.org.

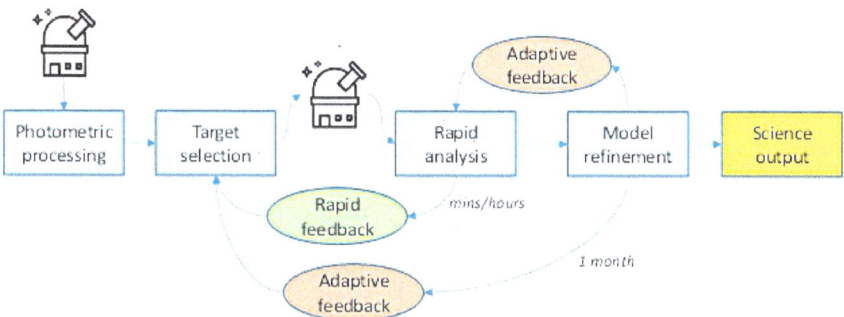

FIGURE 3-1 Use of AI feedback for the Prime Focus Spectrograph at Johns Hopkins University and Princeton University.
SOURCE: Szalay, 2020.

dramatically improving experimental design for using AI to conduct large-scale scientific experiments. One example includes the Next Generation Astronomical Surveys, since observing the spectra of thousands of stars is far more expensive than imaging. For the Subaru Prime Focus Spectrograph project, Johns Hopkins University and Princeton University, with support from the Schmidt Family Foundation, are examining how to use AI feedback from observed targets to improve target selection via reinforcement learning (Figure 3-1).

There is also a need for active curated services to enable scalable data access and analysis. For example, Szalay (2020) compared the evolution of the music industry with the evolution of data science. For the music industry, a traditional long-playing record (LP) or compact disc (CD) has been replaced by download services with apps such as iTunes and online personalized music streaming services. For data science, the equivalent of LP or CD is the process of downloading data (and analyzing them on personal computers); the equivalent of iTunes is the use of data queries to connect to project servers; and the equivalent of music streaming services is that everything will run in the cloud in the next 10 years at an accelerated pace of deployment. According to Szalay (2019), "Algorithms are making the decisions, and soon we will see AI tools setting adaptive choices about survey strategies, including target selection. This may be the beginning of the Fifth Paradigm of Science, where computers decide objectively which experiments will yield the biggest gain in our knowledge."

Looking forward, deep-learning software will become a commodity as users demand AI-ready data sets to conduct AI-driven scientific experiments and build faster proxy simulations (Szalay, 2020). Manual approaches cannot keep up with the sheer volume of astronomical data, and long-term access to data must be FAIR, open, free, and sustainable at the same time (Szalay, 2020). With regard to using automated workflows and data science tools, Szalay (2020) stated that it would be useful to consider active learning for experimental design from planning to execution, the use of AI in analyses with explainable inference, and automated workflows for rapid follow-up of transients.

Barriers to progress include the challenge of ensuring steady, long-term support for preserving irreplaceable data. With increased expectations for open or free access to scientific data, what happens to large, high-value data sets when they are completed? Who should be entrusted with these irreplaceable data and how should the decision be made about what to preserve? He suggested the need for a new trusted intermediary to replace the publishers, akin to a "Smithsonian for Digital Data" with endowments partially supported by the federal government. It would be useful to consider the Smithsonian model for preserving and using long-term data on a 30- to 50-year horizon. Funding agencies could become proactive rather than reactive with a 10-year time delay. Although more are needed, he pointed to an increasing number of trusted, archival data repositories that have developed succession plans for stewardship of their data beyond the life of the specific repository and in some cases are certified using the CoreTrustSeal.[2]

Particle Physics

Modern particle physics involves large collaborations of up to 10,000 researchers structured around very expensive instruments such as the CERN Large Hadron Collider (LHC). Confirming the existence of the Higgs boson, which was announced in 2012, involved combining numerous sources of evidence (Cranmer, 2020). Groups of 10–50 physicists analyzed data from A Toroidal LHC ApparatuS (ATLAS) and Compact Muon Solenoid experiments at the LHC, each group using its own workflow. A technical solution enabling collaborative statistical modeling was developed, and teams were able to combine their data to estimate the probability of generating the actual experimental results given their preexisting theoretical assumptions.

This approach allowed for rapid confirmation of the existence of the Higgs boson and publication of results. However, it was not possible to reproduce the results from what was reported in the original articles, and it took some years to publish the various complex analyses required to do so. Ultimately, physicists began sharing their likelihood scans, allowing others to derive the reported results. Such sharing also allows theorists to test new hypotheses against the experimental results without the need to share the underlying data (Cranmer and Yavin, 2010). The release of full analysis likelihoods from ATLAS reflects this shift in approach (Anthony, 2020).

Looking to the future, active learning and other AI approaches can help power simulation-based inference in physics and potentially in other fields (Cranmer et al., 2020). Researchers faced with inverse problems (i.e., where causal factors are derived from observations, rather than deriving experimental results from

---

[2] See https://www.coretrustseal.org/.

experiments designed to test specific hypotheses or theories) have until recently been limited to labor-intensive approaches relying on expert-generated summary statistics that are not well suited to high-dimensional data (Cranmer et al., 2020). Advances in active learning enable new approaches to drawing inferences from simulators, with the most promising approach in a given case depending on factors such as the inference goals, the dimensionality of the model's parameters, the characteristics of the simulator, and so forth. These capabilities can be combined in workflows that implement various inference algorithms. The new approaches hold the possibility of significantly advancing the productivity of research by allowing experiments to achieve, for example, a given sensitivity using half the data.

Taking full advantage of these new techniques will require rethinking many aspects of the research process (Chen et al., 2019). For example, large facilities might be planned and constructed to advance several scientific goals, rather than a single goal, at the outset, given the broader potential for reusing outputs (Cranmer, 2020). If researchers actively encourage reuse of existing workflows, new theories could be explored more efficiently (Strassler and Thaler, 2019). Embedding support for developing tools that advance simulation-based inference in particle physics within specific projects can speed the emergence of tools and systems that attract broader use.

However, there are barriers to progress in advancing active learning approaches that utilize workflows to physics discovery. For example, sharing data can power these new approaches, but attitudes about data sharing vary widely across subfields (*Nature Physics*, 2019). Given the lack of consensus on data-sharing requirements for publication and whether embargoes on data should be allowed to give data exclusivity to the original research group for months or years, it is not possible for journals to adopt uniform standards. In addition to researcher concerns about being "scooped," there are significant levels of conservatism in senior faculty which, through their control of promotion or tenure and funding decisions, can hamper progress that might be made from the typically more open attitudes of early career researchers (Cranmer, 2020). The link between new techniques and publishing or reporting is especially significant, since publishing is tightly linked to advancement and other rewards that motivate individual researchers and teams.

Materials Science

Materials research "is strongly focused on discovering and producing reliable and economically viable materials, from super alloys to polymer composites, that are used in a vast array of products essential to today's societies and economies" (NASEM, 2019a). The materials discovery process involves the conception of materials on the basis of models, synthesis of new materials, and testing or characterization of these new materials. Traditionally, these steps have been carried out in sequence, with human interventions that have limited throughput. According to Aspuru-Guzik and Persson (2018), it is now possible to close the discovery loop in materials by

> integrating automated robotic machinery with rapid characterization and AI to accelerate the pace of discovery. [This] will unleash a "Moore's law for scientific discovery" that will speed up the discovery of materials at least by a factor of ten—from 20 years to 1 to 2 years. This will catalyze a transition from an Edisonian approach to scientific discovery to an era of inverse design, where the desired property drives the rapid exploration, with the aid of advanced computing and AI, of materials design space and the synthesis of targeted materials.

Service (2019) describes several material science efforts that are pioneering closed-loop experimental workflows. These processes combine automated mixing, preparation, and testing of samples with AI algorithms that evaluate the sample testing results and then decide which materials should be synthesized and tested next. Processes that used to require 9 months for one lab now take 5 days.

McQueen (2020) outlined how ARWs in materials research might evolve. Routine processes will increasingly be automated. Yet, contrary to skepticism in the community, the potential for advances goes beyond optimization or achieving greater throughput and speed. The use of AI in evaluating results and identifying the next set of experimental processes has the potential to qualitatively expand the possibilities of true discovery. Many workflows for analyzing a particular type of characterization exist in isolation, but new value could be unlocked by linking them together. At the same time, it will be necessary to preserve the ability of human researchers to make unexpected observations (McQueen, 2020). Having a "human in the loop" is essential both for making scientific progress and for convincing the materials science community that this is the right way forward.

McQueen identified three interrelated challenges and obstacles to this vision of future materials research: (1) lack of data sharing and access, (2) shortage of researchers who can bridge the gap between materials research experimentalists and those who are developing the necessary software tools, and (3) inertia within the community and resistance to pursuing automated approaches to research powered by AI.

Source data in materials research generally are not shared. This makes it difficult for the community to develop larger-scale shared data resources, as opposed to individual labs relying mainly on the data that they generate themselves. Access to more data is particularly important for identifying the best way to produce a given material by characterizing precisely what was done in one lab and comparing that process with what was done in other labs. Another factor that limits data sharing and reuse is that many instruments used in materials research, such as electron microscopes, generate data only in the proprietary formats unique to each manufacturer.

Regarding human resource needs, there is a gap between the scientific questions being asked and the questions that the data can answer. A translation step is needed, requiring experts with knowledge of data science and material science. However, it is difficult to convince particularly senior researchers of the value of such experts. It will require additional time and effort to arrive at the point where data science and material discovery are no longer regarded as separate activities.

Finally, McQueen said that change is hampered by the fact that current academic reward systems and research assessment approaches generally undervalue creation, sharing, and curation of data or the creation of software that facilitates automated experimentation and AI-facilitated discovery. The peer review process often makes it difficult to get funding for data stewardship or workflow systems innovation. One way to mitigate this problem could be enhancing the collection capabilities of national facilities or multi-institutional consortia.

Kristin Persson (2020a) described one such effort, the Materials Project,[3] which provides open web-based data on known and predicted materials and analysis tools to those searching for novel materials for batteries, solar cells, and computer chips (Persson, 2020b). The project uses high-performance computing within a sophisticated integrated infrastructure consisting of Pymatgen (an open-source Python-based analysis library) and Fireworks (automated open-source workflow software) to determine structural, thermodynamic, electronic, and mechanical properties of most known crystalline inorganic compounds (Hill et al., 2016). A high-level interface to the Materials Application Programming Interface has been built into the Pymatgen analysis library that allows users to programmatically query and analyze large quantities of materials information. To create a more closed-loop workflow, tools are needed to aid in dynamic rerouting, error management, flexibility, and constant communications among domain scientists (Persson, 2020a). Lack of data availability for data-hungry methods limits the use of deep learning.

The Clean Energy Materials Innovation Challenge Expert Workshop in 2017 highlighted the need to develop the materials discovery acceleration platforms that integrate automated robotic machinery with rapid characterization and AI (Aspuru-Guzik and Persson, 2018). Since then, several institutions have been working to advance the field. For example, the University of British Columbia focuses on advanced robotics with synthetic organic chemistry (Hein Lab, 2020), the Air Force Research Laboratory leads research on carbon nanotubes, and the National Institute of Standards and Technology is designing robots that automatically perform experiments recommended by AI with minimal human intervention (Hattrick-Simpers, 2020).

While there are many notable materials data resources (Hill et al., 2016), a centralized repository of information about materials science, similar to the Protein Data Bank that archives information related to the three-dimensional

---

[3] See https://materialsproject.org.

structures of large biological molecules, would be useful (Persson, 2020a). Stable and adequate funding is needed for maintaining and curating data, as data curation emerges as key to progress. Although several federal government initiatives, such as the Material Genome Initiative in 2011 and the Materials Science and Engineering Data Challenge in 2015, have been implemented to encourage the use of publicly available data to model or discover new material properties, additional open data policies are needed to accelerate discovery. There is a need to develop multidisciplinary international teams of scientists and engineers with expertise in chemistry, materials science, advanced computing, robotics, AI, and other relevant disciplines (Aspuru-Guzik and Persson, 2018; Persson, 2020a).

BIOMEDICAL SCIENCE

Biology

According to Murphy (2020), closed-loop experimental systems relying on automation and AI can advance experimental biomedical research. The last several decades have overturned reductionist paradigms in biology, challenging the belief that the functioning of living organisms can best be understood by breaking them down into systems and components that operate according to fixed rules.

For example, conventional understanding of the relationship among DNA, RNA, and proteins is that messenger RNA (mRNA) transcribes information from DNA and then serves as a template for the assembly of a protein that carries out a specific cellular function. However, some exceptions to what was thought to be a fixed relationship among sequence, structure, and function have emerged, such as reverse transcriptase, in which an enzyme (protein) uses an RNA template to generate complementary DNA. Retroviruses use reverse transcription to replicate their genomes. Other exceptions and complexities include transposons (DNA segments that move between different positions in a gene), introns (DNA sequences that are "spliced out" in the formation of the final mRNA), and prions (misfolded proteins that can transmit their shape onto normal proteins of the same type and are hypothesized to be the cause of some diseases). The result is that reductionist paradigms have lost their predictive power in understanding the complexity of cells, tissues, and organisms.

The development of systems biology, which involves building predictive models based on a block of experiments, has been one response. At the workshop, Murphy (2020) argued that systems biology approaches are limited because empirical models cannot be proven correct; they can only be refined and improved over time with more and better data. Given the inherent complexity, absence of fixed rules, and inability to characterize and measure the effect of every biological change on every variable, human understanding is not possible. Thus a new approach is needed in which empirical data are gathered through automated experiments selected by AI.

This approach is being tested in drug discovery (Murphy, 2011). Normally, candidate compounds are screened according to whether and how strongly they affect the biological target associated with a disease. However, choosing the compound with the strongest effect on the target requires additional screening steps to determine whether it acts on other targets that produce toxic side effects. Being able to test all candidate compounds against all targets to find the optimal compound (i.e., strongest effect with no side effects), rather than testing each compound against each target in sequence, will accelerate discovery in this space. Researchers will be able to build empirical predictive models that generate faster, more efficient results.

Use of AI-selected, automated experiments is potentially very powerful. General-purpose tools can be used, in a situation analogous to self-driving cars. The AI is told where to go, not how to get there. As an example, a retrospective test of an approach utilizing active ML yielded promising results (Kangas et al., 2014). The test utilized data from PubChem and involved 177 assays, 133 unique protein targets, 20,000 compounds, and about 1 million experiments. An active learning algorithm was tested against other algorithms to identify compounds affecting the target through simulation of a series of experiments, with the relevant database information hidden from the algorithms. In this analysis, 57 percent of the active compounds were identified by performing 2.5 percent of the possible experiments by the active learning algorithm, compared with 20 percent identified through a traditional approach of building a model for each target. More recently, a prospective study was undertaken in Murphy's laboratory at Carnegie Mellon University using laboratory automation to execute experiments selected by AI to model complex phenotypes (Naik et al., 2016). There was no human in the loop in this case. By performing 28 percent of the possible experiments, the model was 92 percent accurate, about 40 percent more accurate than what was achievable using random selection.

Murphy identified several barriers to progress, beginning with a shortage of experts who can bridge the gap between disciplinary and laboratory automation expertise. Autonomous laboratory automation and standards are also needed. Several companies are now providing services in this space, such as Emerald Cloud Lab[4] and Strateos,[5] that perform experiments on demand. Real-time primary data sharing from individual labs, and the infrastructure to do so, are also needed. National research resources that execute particular types of automated experiments on request—analogous to what exists in astronomy and physics—are also needed. The ability to test alternative approaches through such facilities will enable cost-sensitive, proactive learning.

Finally, researcher training will need to change. Currently, graduate students in biology are taught to pipette and manipulate experimental conditions by hand.

---

[4] See https://www.emeraldcloudlab.com.
[5] See https://www.strateos.com.

In the future, it will be more important that researchers are able to choose the overall goals of their research and design a campaign using diverse technologies to reach the goal. Humans will also need to invent new measurement technologies, and to develop and improve AI methods.

Biochemistry

Chemical synthesis is a two-centuries-old empirical science and is the bottleneck in the optimization of drug discovery, a process that typically takes years (Cernak, 2020). A tighter coupling between chemical synthesis and data science would accelerate this process. The COVID-19 pandemic illustrates the urgent need for rapid development and testing of novel medications in times of viral outbreaks.

Cernak (2020) identified high-throughput synthesis and screening as areas ripe for new advances. Small molecules used in pharmaceutical applications, including antiviral medications such as those needed to treat COVID-19, are difficult and time-consuming to manufacture and screen. They require multistep reactions to synthesize. The objective is to use this design–make–test cycle to identify molecules that combine multiple properties such as permeability, solubility, selectivity, stability, potency, and synthetic accessibility (see Figure 3-2).

Manufacturing, in particular, is a critical bottleneck in this cycle. Researchers are working to accelerate chemical synthesis through automated reaction equipment and in-line/in-situ analysis tools. For example, Timothy Cernak's lab at the University of Michigan is able to perform 1,500 experiments rapidly through nanoscale synthesis using robotics (Cernak, 2020). This provides a systematic snapshot of the reaction landscape. Reaction enumeration, which involves transforming the chemical reaction into a machine-readable form, such as a graph, is another important step. The ultimate aim is to automate the process of retrosynthetic analysis through data science (Cernak, 2016).

An example of efforts along these lines is the Center for Computer-Assisted Synthesis,[6] supported by the National Science Foundation. Companies are also working within this new paradigm for drug discovery. Insitro[7] seeks to "leverage the tools of modern biology to generate high-quality, large data sets optimized for machine learning." X-chem is a premier DNA-encoded library technology company that announced a spinout company called ZebiAI[8] to work in partnership with an ML group at Google.

Barriers to progress within the community include resistance and lack of large-scale resources to enable automation. Regarding the former, chemical synthesis has traditionally been considered something of an artisan process, and so

---

[6] See https://ccas.nd.edu/.
[7] See http://insitro.com/about.
[8] See https://www.zebiai.com/about-us.

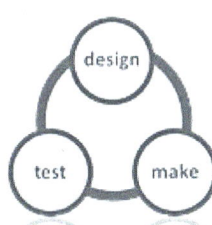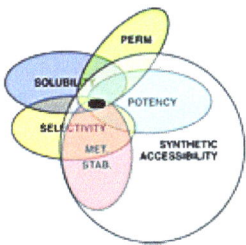

FIGURE 3-2 Chemical space exploration.
SOURCE: Cernak, 2020.

gaining acceptance for automating important components of the process is difficult. The challenge is to "teach data science to organic chemists" (Cernak, 2020). The lack of large-scale resources for automation, such as sufficiently characterized data, is another obstacle. Cernak's lab generates its own data since much of the data openly available in the literature may be noisy or lack sufficient metadata for machine readability. Proprietary database software systems such as Reaxys or SciFinder limit the volume of entries that can be accessed without buying access to the entire database, which academic labs might find prohibitively expensive.

HEALTH AND ENVIRONMENT

Epidemiology

The pandemic induced by COVID-19 has touched all aspects of society. It has also created opportunities to assess the capabilities of modern scientific workflows and to innovate new paradigms. In this respect, it is important to recognize that the appearance of COVID-19 coincides with the rise of cheap ubiquitous sensors and the big data revolution, such that actions are increasingly collected, stored, and made accessible for analytic investigation in a digitized form. By combining modern workflow systems with large amounts of personal data, drawn from a wide variety of domains, there is an expectation for breakthroughs in public health policy and assessment, rapid refinement of guidelines for clinical care, repurpose of known drugs for treatment, and crafting of novel vaccines.

Many examples of rapidly organized scientific endeavors associated with COVID-19 have emerged. For instance, various consortia, such as the National COVID Cohort Collaborative (N3C) funded by the National Center for Advancing Translational Sciences of the National Institutes of Health, have been formed to collect and make accessible clinical data about individuals infected with COVID-19 (NCATS, 2021; Rubin, 2020). The Global Platform for the Analysis of SARS-CoV-2 Data was launched by an international collaboration on the Galaxy platform (Galaxy, 2021b; Maier et al., 2021). The Virus Outbreak Data Network is a coordinated effort to allow observational patient data from various heterogeneous information systems to be made FAIR and machine actionable (Queralt-Rosinach et al., 2021). From a discovery perspective, AI-driven workflows have been applied to sift through large lists of drugs to search for those that could interfere with mechanisms by which COVID-19 infects cells. It was further shown that workflow automation could be combined with techniques from network science to search for drugs that would interact with groups of proteins associated with the disease.

These efforts have generated promising results. For example, a 2021 article reports on a team that used N3C to assemble the largest nationally representative cohort of patients with cancer and COVID-19 and are using this cohort to better understand the effects of COVID-19 on cancer outcomes and improve treatments (Sharafeldin et al., 2021). Another team used N3C to compare the effects of alternative treatments for hyperglycemia in SARS-CoV-2–positive adults (Kahkoska et al., 2021).

However, the extent to which these breakthroughs could lead to actual treatments and, more generally, to solutions to pandemics is dependent on a number of factors. This is because the search for answers is not simply a matter of collecting vast amounts of data and subjecting them to continuous automated analysis. Rather, it requires addressing a complex combination of technical, legal, and social affairs that need to be resolved in order to be successful. For instance, in the context of clinical care, the gold standard by which care routines, interventions, and treatments are assessed is the randomized clinical trial, which reduces sample bias. Yet the data currently being collected in the clinical domain with respect to COVID-19 are not the outcome of a randomized clinical trial and are thus subject to certain biases. For example, patients that present to health care facilities generally only present when they have symptoms that are recognizable and have the ability to access a testing facility (whether it be at a large medical center or a smaller stand-alone testing center).

One of the challenges in attempting to rapidly respond to an emergency is the potential to shirk provenance. In scientific investigation, particularly when the investigation is in the general public interest, it is critical to ensure that the data collected for the investigation are correct. This does not mean that the data must be made publicly available for all to inspect and scrutinize, particularly when there are competing interests (e.g., privacy concerns for patients who did not consent to the public disclosure of their identities), but that there must be proof that the data are not fabricated and have not been doctored. Unfortunately, in the case of COVID-19, there have been clear cases where the rush to collect and analyze data has led to questionable results.

These cautionary tales are not indications that the system for data collection, management, and analysis is fundamentally broken. Rather, it provides an illustration of what can go wrong when a system is hastily erected. As we work toward the development of automated methods for pandemic detection and subsequent mitigation, it will be critical to ensure that the data have clear provenance, that the workflows and analytics conducted within them are verifiable, and that plans for data sharing, access, and accountability for abuse and misuse of the data, as well as findings from such workflows, are established from the outset. In this respect, it is critical to continue to create common data models, promote approaches to mitigate bias in data collection and analysis, and support the infrastructure necessary to support—if not in real time, at least in near real time—scientific investigation that leads to guidance on how best to mitigate public health threats.

Climate Science

Climate change is upon us. Even with mitigation strategies, it requires adaptation to a new normal, but what that new normal will be remains unclear. Some large-scale features of climate change, such as the more rapid warming of continents relative to oceans, are well understood and well simulated by models. But projections of climate change and the associated risks, for example, of flooding, wildfires, and high-impact weather events, continue to be marred by large and poorly quantified uncertainties that hamper informed decision making and make it more difficult to adapt proactively and effectively (e.g., Schneider et al., 2017b; Global Commission on Adaptation, 2019; Hill, 2021; IPCC, 2021).

The principal source of uncertainties in climate projections is small-scale processes, such as the turbulent motions and microphysical processes controlling clouds, turbulent mixing in the oceans, or small-scale frictional or rheological processes at the base of or within ice sheets. These small-scale processes are too costly to compute globally, yet are crucial for modeling climate (NASEM, 2012). High-resolution simulations in limited areas have been used for some time to inform reduced-order models of clouds and of atmosphere and ocean turbulence, which are used in global models. Typically, the high-resolution simulations are carried out in a few locations, at selected times of year, for example, at locations and times when data from field campaigns are available (e.g., Siebesma et al., 2003; Stevens et al., 2005). This approach has provided a limited sample of computationally generated data.

The workflow for carrying out high-resolution simulations, calibrating reduced-order models of small-scale processes, and quantifying their uncertainties can now be automated and accelerated by orders of magnitude. Targeted high-resolution simulations can be spun off from a global climate model on the fly, wherever and whenever needed, to provide detailed information that the global model on its own is unable to provide (Schneider et al., 2017a). Spinning off such high-resolution simulations lends itself well to distributed computing approaches and hence is very scalable. Instead of exploiting such high-resolution simulations in only a few locations and at selected times of year, experimental-design algorithms can be used to optimize the placement of high-resolution simulations to be maximally informative about small-scale processes in global models. Tools from data assimilation, pioneered in weather forecasting (Kalnay, 2003), can be combined with newer techniques from ML to accelerate learning about small-scale process models in computationally expensive climate models (Cleary et al., 2021). More data-hungry, deep-learning methods can also be used in place of process-based, reduced-order models (e.g., Rasp et al., 2018; Yuval and O'Gorman, 2020).

Additionally, similar tools can be used for a global climate model to learn automatically from the plethora of global climate data that are now available, be they from space, the ground, or autonomous ocean vehicles (Schneider et al., 2017a). Conversely, a climate model that learns automatically from observational data can help determine the value that new observing systems would provide, for example, in terms of reduced uncertainties in climate projections. Gaining such insights "in observing system simulation experiments (OSSEs) is increasingly required before the acquisition of new observing systems (e.g., as part of the U.S. Weather Research and Forecasting Innovation Act of 2017)" (Schneider et al., 2017a). More generally, automating the workflow for learning from observational data makes it possible to quantitatively pose the experimental design question about what kind of observations would be maximally informative to reduce climate model uncertainties further. Figure 3-3 illustrates a model–learn loop for climate modeling.

Weather forecasts have made great strides over the past decades, thanks to improvements in the automated assimilation of observations (Bauer et al., 2015). Climate projections can now advance similarly, by simultaneously harnessing observations and data generated computationally in high-resolution simulations. The acceleration in the rate of improvement of climate models has the potential to lead to a qualitative leap in the accuracy of climate projections.

Unlike in other fields, climate sciences and weather forecasting data have been open, accessible, and widely shared worldwide, going back to global data-sharing frameworks developed in the 1950s. This openness facilitates the development of ARWs and quality control of data by cross-checking among

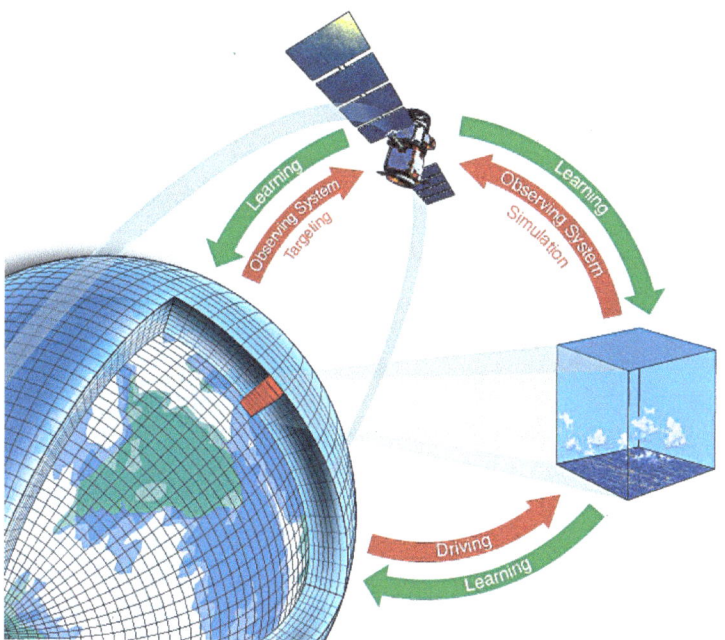

FIGURE 3-3 A model–learn loop for climate modeling.
SOURCE: Schneider et al., 2017a.

disparate data sets. However, the openness and the benefits that accrue from it are now under threat because government agencies are beginning to purchase data from commercial providers under restrictive licenses.

Wildfire Detection

There is an urgent need for better modeling of a range of environmental hazards, including societal impacts, and innovative approaches combining advanced computing, remote sensing, data science, and the social sciences hold the promise of mitigating hazards and improving responses. In pursuing these advances, "there are still challenges and opportunities in integration of the scientific discoveries and data-driven methods for detecting hazards with the advances in technology and computing in a way that provides and enables different modalities of sensing and computing" (Altintas, 2019). The National Science Foundation–funded WIFIRE project[9] created an integrated system with services for wildfire monitoring, simulation, and response, providing an end-to-end management infrastructure

---

[9] See wifire.ucsd.edu.

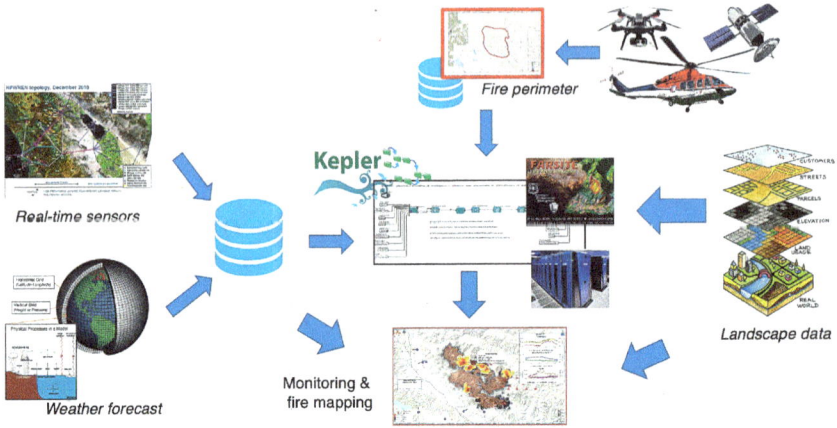

FIGURE 3-4 Dynamic data-driven fire modeling workflows in WIFIRE.
SOURCE: Ilkay Altintas.

from data sensing and collection to modeling efforts using a continuum of computing methods that integrate edge, cloud, and high-performance computing (see Figure 3-4). The multidisciplinary team formed significant partnerships with data providers, science communities, fire practitioners, and government organizations to solve problems in wildland fire. The WIFIRE project is used for data-driven knowledge and decision support by a wide range of public- and private-sector users for scientific, municipal, and educational purposes.

The integrating factor in WIFIRE is the use of scientific workflow engines as a part of the cyberinfrastructure to bring together steps involving AI techniques on data from networked observations (e.g., heterogeneous satellite data and real-time remote sensor data and computational techniques in signal processing, visualization, fire simulation, and data assimilation) to provide a scalable, repeatable, and customizable closed-loop capability.

In addition, WIFIRE represents a common theme of applications using ML on top of nontraditional hardware and making use of the new processors that have emerged in recent years, including graphics processing units, field programmable gate arrays, and edge accelerators. The common theme of these applications, integrating AI workloads, is "their need to run in specialized environments for reasons such as the on-demand or 24x7 nature of the tasks they are performing, and difficulties regarding their portability, latency, privacy, and performance optimization" (Altintas, 2020a). Moreover, there is a need for composable systems where these applications are integrated with high-performance computing or high-throughput computing tasks.

A typical example of these applications is the role of real-time edge processing and the use of ML and big data in wildfire behavior modeling applications

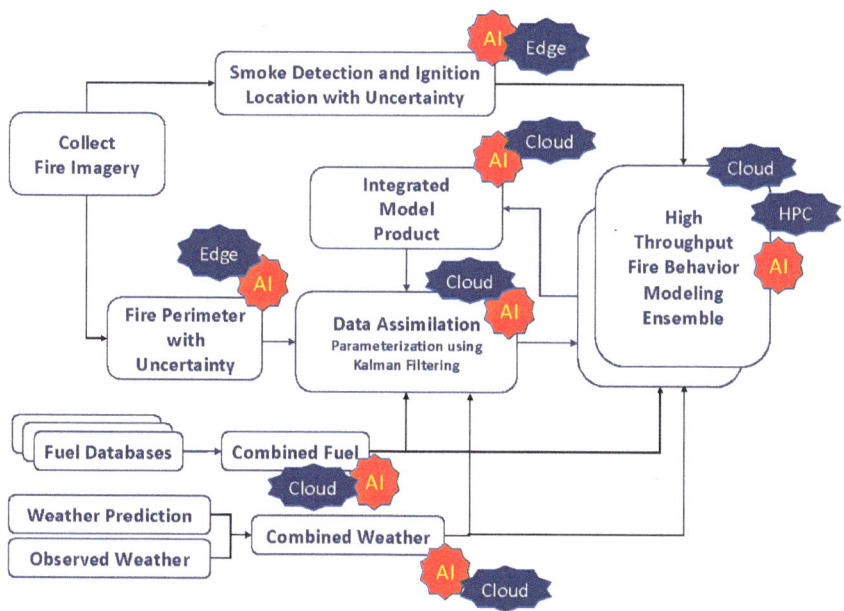

FIGURE 3-5 Scenario for AI-integrated fire modeling at the continuum of computing.
SOURCE: Ilkay Altintas.

within the WIFIRE cyberinfrastructure. WIFIRE's dynamic data-driven fire modeling workflows depend on continuous adjustment of fire modeling ensembles using observations on fire perimeters generated from imagery captured by a variety of data sources including ground-based cameras, satellites, and aircraft. Typically, perimeter generation is performed in a big data and/or edge computing environment while fire modeling is performed in an HPC or HTC environment, depending on which fire modeling codes are executed (Altintas, 2020b). Figure 3-5 illustrates an integrated workflow with the various steps marked with the execution infrastructure required.

Learning from imaging to decide on parameterization of simulations and combining this learning with the insights gained from analyzing the time-series output of simulations promise many benefits ranging from shorter runtimes and time to discovery, to decision support for urgent applications, and less energy usage for computing.

Currently, there are many new tools and technologies (e.g., for scheduling and resource monitoring) to scale and manage the components of these application workflows homogeneously. However, these application workflows are composed of steps that require a heterogeneous cyberinfrastructure ecosystem that involves dynamic resource management and coordination. Current cyberinfrastructure lacks an integrated environment that can pull data from a number of resource monitoring tools and turn these federated data into predictive intelligence needed to steer application workflows in a dynamically scalable and data-driven fashion at the time of workflow execution.

DIGITAL HUMANITIES

Digital humanities make use of computational tools to conduct textual search, visual analytics, data mining, statistics, and natural language processing (Biemann et al., 2014). Similar to applications in science and engineering,

> the processing of large data sets with appropriate software opens up novel and fruitful approaches to questions in the "traditional" humanities. Thus the computational paradigm has the power to transform them. One reason is that this kind of processing opens the way to new research questions in the humanities and especially for different methodologies for answering them. Further, it allows for analyzing much larger amounts of data in a quantitative and automated fashion. (Biemann et al., 2014, pp. 80–81)

The past decade or so has seen development of capacity and expertise in the creation of workflows for the humanities, which require "abstract representations of research phases, taxonomies of scholarly activities, in conceptual frameworks, or in scholarly ontologies" (Koolen et al., 2020). Subdisciplines have both the benefit and the challenge of ever-growing amounts of data, the quantity of which no human (or group of humans) could digest in analog format. Examples of digital repositories to collect and make use of data include the Open Islamicate Texts Initiative[10] and the Chinese Text Project,[11] both of which contain 1 billion words and are growing.

One notable example of a leader in the field is the King's College London Digital Lab.[12] A team of about 15 research software engineers and other experts support the research conducted by other academic departments, including cultural heritage, history, and archeology, as well as outside organizations. Examples range from a project to determine the origins of the Gough Map (the first known map of the British Isles) to archiving and making accessible the roughly 100 different digital research projects that the lab inherited upon its creation in 2015 (Smithies et al., 2019). Another relevant project is the Machine Learning for Music project, "a community of composers, musicians, and audiovisual artists, exploring the creative use of emerging Artificial Intelligence and Machine Learning technologies in music."[13]

---

[10] See https://iti-corpus.github.io.
[11] See https://ctext.org.
[12] See http://www.kdl.kcl.ac.uk.
[13] See https://www.unsupervised.uk/about.

As the field has emerged, the Alliance of Digital Humanities Organizations (ADHO)[14] formed in 2005. It now comprises 10 professional societies worldwide, including the Association of Computers and the Humanities, based in the United States, and the European Association for Digital Humanities, among others. Its mission is to promote and support "digital research and teaching across all arts and humanities disciplines, acting as a community-based advisory force, and supporting excellence in research, publication, collaboration, and training." ADHO members publish peer-reviewed journals (such as *DSH: Digital Scholarship in the Humanities*), contribute to reference books developed for the field, and host discussion forums and conferences.

Efforts to integrate digital arts and humanities across Europe include two initiatives: DARIAH-EU (Digital Research Infrastructure for the Arts and Humanities)[15] and the Common Language Resources and Technology Infrastructure.[16] In the United States, the Office of Digital Humanities within the National Endowment for the Humanities offers grants to digital projects, many of which produce white papers for further knowledge sharing. These efforts could contribute to the integration of resources necessary to apply next-generation workflows to humanities research.

A case study presented at the workshop involved philology to show the potential and challenges of ARWs in the humanities. Philology involves linguistic or textual records to reconstruct "anything that happened in the human mind or outside in the world, from understanding a Cuneiform text from the ancient world to how Iranian, Farsi-language newspapers are reporting on the coronavirus today" (Crane, 2020). While philology predates the use of digital tools by centuries, in recent years, computational approaches have had the potential to transform what researchers can learn through textual records.

In addition to the sheer number of words in any one language, researchers are faced with the multitude of languages, ancient and modern, through which people have communicated ideas over the centuries. Crane (2020) acknowledged that a principal challenge is not "1 billion words of English language newspapers but 1 million words of poetry in 100 languages," which calls for networks of deep annotation, that is, machine-actionable representations of fine-grained judgments about a source. The ability to provide precise citations across texts is also aided by digital tools, a need that also exists in the sciences. Tools can now reference any word, symbol, or element in any surviving text-bearing object.

AI-enhanced tools are also allowing researchers to see patterns never visible before. For example, the Viral Texts Project[17] is mapping networks of the reprinting of articles in 19th-century newspapers and magazines. Text is generated from scans of newspapers, with their millions of words, even in cases where the texts

---

[14] See https://adho.org/.
[15] See https://www.dariah.eu/.
[16] See https://www.eudat.eu/use-cases/common-language-resources-and-technology-infrastructure.
[17] See https://viraltexts.org/.

are degraded and barely legible for humans to read. Thus, researchers can trace the spread of ideas, such as abolition of slavery, to understand how certain ideas became "viral" in a historic context.

The vast amount of material written in Latin pre-1750 represents a great understudied collection of materials that AI can address, according to Crane. Through that time, Latin was the majority language of scholarship in most European countries, no matter the everyday spoken language, and was used in dissertations, scientific publications, and other materials. Millions of documents are lying unread in libraries and archives. Digitizing the materials, and then setting up workflows to allow for search and extraction can provide clues to the development of European thought. A particular quotation about a concept used in one book can be searched for in 3 million other books to understand how it is used in different times and contexts. Similarly, an amorphous concept such as "honor" can be traced to see the meaning placed on it by different cultures.

Similar to the sciences and engineering, ARWs in the humanities result in a hybrid environment that integrates human feedback and contributions with ongoing automated analysis of linguistic sources. The machines can mine data, but a human in the loop must provide the training material that drives the artificial intelligence systems and corrects raw material that gets fed back into the system. While ARWs can mine data, suggest patterns, and point to new directions, humans are a critical part of the ecosystem. According to Crane, the digital humanities are at a pivotal moment, as a new generation of scholars engages in research. "There is a gap between traditional research and these new tools," he said at the committee's March workshop. "What faculty are trained to do and what graduate students are equipped to do is different."

SOCIAL AND BEHAVIORAL SCIENCES

The factors that are driving and facilitating the implementation of ARWs in the social and behavioral sciences (including economics, sociology, political science, and psychology) are broadly similar to those operating in other domains. These include the rapidly growing ability to access large amounts of high-quality data—particularly streaming data—at relatively low cost, the advance of ML tools for deriving insights from those data, and the urgency and importance of the questions that can be addressed using these approaches. At the same time, researchers in the social and behavioral sciences face some of the same barriers to the advance of ARWs as those faced in other domains, as well as some distinct issues.

Traditionally, researchers in the social and behavioral sciences have mainly worked with small or medium-size data sets, such as survey data collected by researchers themselves, or data generated by government statistical agencies on a scale allowing their downloading and analysis by the researchers themselves using standard statistical software (Turner and Lambert, 2014). The sorts of questions addressed include how tax policy changes affect gross domestic product growth and income distribution and how the policy positions of political parties affect voter preferences. Increasingly, statistical agencies such as the Census Bureau, Bureau of Labor Statistics, and Social Security Administration are making larger data sets available. Researchers are also able to access data from private companies such as Google, Amazon, and credit card companies through special agreements (Einav and Levin, 2014).

There is growing recognition of the potential for new data resources to deliver real-time, actionable information. The Foundations for Evidence-Based Policymaking Act (Public Law 115-435) received bipartisan support when it passed Congress in 2019. It requires "agency data to be accessible and requires agencies to plan to develop statistical evidence to support policy making". As Julia Lane commented at the committee's workshop, "We are at a golden moment in which lots of people care. . . . Data sits at the core of what federal agencies, and state and local agencies, are asked to do."

The following examples illustrate how new data resources and advanced analytics are being applied in the social and behavioral sciences. While these approaches do not constitute ARWs as such, they involve innovative uses of information technology that will likely facilitate the use of AI in research and future implementation of ARWs. The first example is Opportunity Insights, directed by economist Raj Chetty at Harvard University. This project identifies barriers to economic and social mobility in the United States. An Opportunity Atlas considers more than a dozen neighborhood-level characteristics, drawn from multiple data sets, to answer a basic but critical question: Which neighborhoods in America offer children the best chance to rise out of poverty? Opportunity Insights not only shares its publications on its website,[18] but also its data sets and replication codes, as well as the lectures, slides, and other materials used in Chetty's course on using data to solve economic and social problems.

Another example is urban informatics, which harnesses public data to improve services and empower communities. In one application, Daniel O'Brien of Northeastern University analyzed a data set of calls to Boston's 311 system (in which citizens can report non-emergency needs to city officials) to draw conclusions for city leaders and community groups (O'Brien et al., 2017). He has emphasized the millions of interactions captured in digital records, including birth, marriage, and death certificates; school test scores; and building permits, to name a few.

Some social and behavioral sciences researchers are explicitly applying ARWs in their work. Wu et al. (2020) have developed a workflow that they suggest has broader applicability in order to study the impact of Chinese migration patterns, specifically on children who are left behind in China's provinces while their parents work elsewhere. The workflow allows for feature selection criteria through ML approaches not bound by the researchers' knowledge. In addition,

---

[18] See https://opportunityinsights.org.

the ARW proposed by the authors more fully deals with dynamically changing, complex variables than standard regression methods. Similarly, Cedeno-Mieles et al. (2020) proposed a system for networked social science experiments that automates the steps involved in analyzing experimental data, building models to capture the behavior of human subjects, and providing data to test hypotheses. Finally, Bergmann et al. (2021) have developed a tutorial to help economists use the Snakemake tool for managing reproducible data analysis workflows.

As in other disciplines, the amount of data becoming available to social and behavioral scientists presents challenges related to the size of the data sets, as well as ensuring the integrity of identifiable personal information, reproducibility of results, and archiving. According to Lane (2020), the massive scale and dispersed location of data sets, as well as lack of clarity about the level of expertise behind their production, limit their potential contribution and broader use. To overcome this challenge, the Coleridge Initiative[19] (of which Lane is a cofounder) has as its goal to "change the empirical foundation of social science, statistical and public agencies in the United States and transform understanding of how our society works." Among other efforts, the initiative sponsored a Rich Context competition during 2018 and 2019, with a focus on the social sciences, that aimed "to implement AI to automatically extract metadata from research—identifying data sets in publications, authors, and experts, and methodologies used" (Lane et al., 2020). Twenty teams from around the world participated in a first phase of the competition, and four finalists (one from the United States, one from South Korea, and two from Germany) participated in a second phase in 2019 (Lane et al., 2020). The great challenges of our time are human in nature—climate change, terrorism, overuse of natural resources, the nature of work, and so on—and these require robust social science to understand their causes and consequences. Effective use of data for social science research depends on understanding how data sets have been produced and how they have been used in previous works. The opportunity at hand is to leverage ML advances to create feedback loops among the entities involved: researchers, data sets, data publishers, publications, and so on (Nathan, 2020).

Although social and behavioral scientists have embraced many new technologies, some have not taken full advantage of automation but continue to rely on manual and ad hoc techniques (Yarkoni et al., 2021). The five core investments in automation that Yarkoni et al. (2021) recommend extend to other disciplines, but they presented their case through a social science lens:

- *Standardization:* Recognizing that a single standard is neither possible nor desirable in the social sciences, they suggest "an efficient way to ensure commensurability while encouraging innovation is to develop meta-standards that support common access to different standards."

---

[19] See https://coleridgeinitiative.org/.

- *Data access:* Given the nature of the field, data on human subjects provide critical insights—but the reuse of those data are often restricted. Automated data extraction may facilitate both safer and increased access to data.
- *Search and discoverability:* Adapting existing techniques from other fields can help social scientists in overcoming the challenge of finding the relevant findings among an overwhelming number of articles and data sets.
- *Claim validation:* Automation related to enumeration of statistical assumptions, sensitivity analysis to assess assumptions that support a result, and creation of checklists of researchers to verify in-progress work.
- *Automated insights:* In keeping with the focus of this study on ARWs to contribute to new discoveries, the fifth and final area particularly merits attention. As they wrote, "Perhaps the most tantalizing opportunities for automation-related advancement of social science lie in the area of scientific discovery and insight generation." Examples include signal discovery and hypothesis generation, automated meta-analysis, and development of reasoning systems or inference engines.

The social and behavioral sciences have several existing practices and institutions that can help facilitate the development and implementation of ARWs. For example, there are established organizations charged with data stewardship and related training such as the Inter-university Consortium for Political and Social Research and the National Opinion Research Center. Finally, the *American Journal of Political Science* requires that the data supporting published articles undergo an independent verification process.

4

Automatic Research Workflows and Implications for Advancing Research Integrity, Reproducibility, and Dissemination

Researchers across and beyond the disciplines described in Chapter 3 share challenges related to how they plan, conduct, and disseminate their work. They face pressures to secure and sustain funding, collaborate with others, and communicate results expeditiously and accurately. They need to do this in structures that range from small, vertically structured labs involving supervision and mentorship to huge dispersed networks across institutions with no formal lines of authority. In this context comes another overlay: the need to conduct experiments in a manner that allows others not only to understand the findings, but also to have access to and use of the data and methods to arrive at those findings.

During the workshop and in discussions, the committee considered how automated research workflows (ARWs) contribute to these crosscutting research issues. Cyberinfrastructure-enabled research (NSF, 2007) is now essential, but, as was generally agreed in the workshop, its use can also introduce inaccuracies and skewed results if not well understood or allowed to self-perpetuate without human oversight. This chapter covers the relationship between ARWs and issues related to integrity, reproducibility and replicability, and dissemination of research.

INTEGRITY

The increasing complexity of research and the associated data it produces, combined with ever-increasing hypercompetition among researchers in many fields, has given rise to a number of persistent research integrity challenges (*Nature*, 2017; Bucci, 2018). These challenges range from detrimental research practices, such as authorship misrepresentation and inappropriate use of statistical analysis, to research misconduct in the form of data falsification and fabrication.

When exposed, these cases not only have negative repercussions on the individual researcher and/or research group, but also can lead to a lack of trust in scholarly research by the broader community and the public.

The increasing concern across the research ecosystem has led to major reports and recommendations on how to improve the integrity of research (e.g., ALLEA, 2017; NASEM, 2017; WCRI, 2019). Interestingly, a report by the European Commission's Open Science Policy Platform reveals that while some research enterprise stakeholders such as research institutions and learned societies believe that significant progress has been made in addressing integrity issues, stakeholders such as research funders, libraries, publishers, and other organizations that disseminate research believe that much still needs to be done (EC, 2020).

Technological advances may create new ways for researchers to manipulate results, as well as lead to development of new tools to detect mistakes and misbehavior. As noted in the National Academies' 2017 report *Fostering Integrity in Research*:

> In theory, if not always in practice, all the data contributing to a research result can now be stored electronically and communicated to interested researchers. However, this trend toward greater transparency has created tasks and responsibilities for researchers and the research enterprise that did not previously exist, such as creating, documenting, storing, and sharing scientific software and immense databases and providing guidance in the use of these new digital objects (NASEM, 2017, pp. 47–48).

This significant increase in both opportunity and complexity can create challenges in ensuring that researchers at all career and seniority levels receive adequate and regular training in good research practices, applying this knowledge not only in their own work but also in undertaking peer review and other evaluation of the work of others. Furthermore, the increased complexity of much research can make it challenging (in time and effort) to adequately capture and report all elements of an experiment, and then imposes similar challenges for peer reviewers in adequately assessing all this detailed information, with few incentives to authors or reviewers to undertake this effort.

ARWs provide a significant opportunity to address these issues and hence enhance research integrity by

- Enabling automated capture and retention of data and their associated metadata in cyberinfrastructure deployed across the research life cycle.
- Better documentation and reporting of the details of the methods, increasing the ability of other researchers to scrutinize the work and potentially reducing the possibility of data being falsified or results being selectively reported.
- Accounting for limited sample size to reduce *p*-hacking if uncertainty quantification is incorporated into the workflow.

- Providing much greater transparency in processes, which can highlight uncertainties.
- Supporting the comparison of findings to highlight inconsistencies and outliers and to provide transparency to determine whether these elements have been "cleaned up" for the final presentation of the data.

But workflows do not guarantee transparency or, more broadly, integrity. To achieve and maintain integrity, it is crucial to involve humans in the process. An overreliance on naive machine learning (ML) can in itself introduce *p*-hacking and other errors, warned Rebecca Nugent at the March 2020 workshop (Nugent, 2020), often from the data handlers' lack of training or simply being overwhelmed by the amount of data available. Simple bugs and errors in software may result in mistakes in reported results. In cases in which people rely on automated techniques to help them identify *something* that might be useful, the result may be HARKing or *p*-hacking.[1] Caution needs to be exercised around the biases inherent in algorithms, which can easily self-perpetuate and lead to suggested correlations that make no sense. Closed algorithms can be difficult for peer reviewers to assess and spot where such biases may have occurred that may have influenced the final results and conclusions. In addition, although data-driven models have the ability to self-learn and adapt, they sometimes do so blindly. An iterative loop is needed, in which algorithms are adapted based on domain knowledge: domain knowledge is refined based on what the algorithms learn, the algorithms are improved to become more robust, and so on. A new generation of artificial intelligence (AI) algorithms, an "AI 3.0" as it has been called, would better integrate domain-based models with data-driven models (Vidal, 2020).

There is also a need for much higher quality data on which such algorithms might work, to avoid garbage in, garbage out. Data curation becomes an increasingly important task to ensure that small errors in a data set do not get amplified by the use of automated processes that then use that output to inform the next experiment. The task is complex. It involves assessment of which data sets to prioritize for the considerable effort involved in curation, as well as training, incentives to prioritize the effort above other tasks such as conducting further experiments that might lead to further publications, and funding to support the extra effort and cost involved. Several studies have suggested that data stewards can provide research teams with this expertise (e.g., Scholtens et al., 2019), and Barend Mons has called for 5 percent of research funding to be dedicated to ensuring that data are reusable (Mons, 2020).

---

[1] HARKING, or "Hypothesizing after the Results are Known," was so-named by psychologist Norman Kerr in 1998. *P*-hacking refers to manipulating statistical data to show that a result is more significant than it is. In 2016, the American Statistical Association issued a statement to address the proliferation of incorrect use of *p* values (Wasserstein and Lazar, 2016).

REPRODUCIBILITY AND REPLICABILITY

Reproducibility and replicability are crucial elements to ensuring the integrity of scientific processes and trust and reliability of new discoveries on which the next discoveries are built.[2] Attempts to reproduce and replicate work require the original researchers to transparently share their underlying data and the associated methods, as well as the estimation, characterization, and reporting of uncertainty (NASEM, 2019b, p. 6). Given the size and complexities of these investigations, manual capture is challenging and sometimes not possible.

The growing ubiquity and complexity of computation in the research process across many disciplines presents additional challenges to independently reproducing results. Examples of these challenges include the use of nonpublic data and code in research, the costs of retrofitting long-standing research projects with tools that automatically capture logs of computational decisions, and incomplete information about the computing environment where the research was originally performed (NASEM, 2019b). Also, research that utilizes high-performance algorithms and parallel processing may produce different numerical outputs from the same input data on different runs, with the output being "understood as an approximation to the correct value within a certain accepted uncertainty" (NASEM 2019b).

ARWs can enhance research reproducibility and replicability by

- Capturing the provenance of the results, and the data and models on which they are built, thereby supporting the more accurate rerunning of processes. This can include capturing more detail of the methods than might be achievable manually but which might have a material impact on the ability to replicate or reproduce a finding.
- Providing simpler and more efficient approaches to the sharing of the research processes.
- Increasing efficiencies and eliminating a potential source of errors in onboarding new research team members, and better supporting knowledge transfer as research teams change.
- Supporting broader and more stringent review and validation of findings, including through formal peer review during publication as well as by the broader research community when data and associated methods, materials, and code are published.
- Supporting full transparency of the research process to reduce the integrity issues mentioned above, whether deliberately or through poor research practice (e.g., removing outliers, selective reporting, image manipulation, etc.).

---

[2] As defined in NASEM (2019b), reproducibility involves "obtaining consistent results using the same input data; computational steps, methods, and code; and conditions of analysis." Replicability involves "obtaining consistent results across studies aimed at answering the same scientific question, each of which has obtained its own data."

- Capturing validation, seamless integration, and repeatability in team science approaches, which are increasingly the reality in attempting to solve complex problems.[3]

Even without reproducibility and replicability set as primary goals, workflows help achieve these goals by virtue of the features they offer users. As the previous National Academies committee on reproducibility and replicability highlighted, workflow systems such as Nextflow, Galaxy, and Snakemake in the life sciences, Chimera in physics, and the Open Science Framework in psychology (as illustrative examples only) can link research results to the computational processes that derived them (NASEM, 2019b). Blockchain can potentially lock in protocols and outputs so that it is clear that nothing has been interfered with, whether deliberately or through poor research practices.

Capturing provenance is considered a benefit of most workflows, but they are not a panacea. As pointed out in a review of workflows' role in reproducibility, computational limitations include interoperability gaps, use of third-party services that may have reliability issues, and lack of central repositories (Cohen-Boulakia et al., 2017). Even when code and data are available, it may be difficult or impossible to reproduce work when the particular computational environment in which data have been processed, including specific versions of language and systems libraries, cannot be recreated. Capturing provenance and ensuring reproducibility in research that involves issues such as complex computation, rapid or continuous adjustments to analytical processes, and large amounts of streaming data will continue to raise challenges for designing and implementing ARWs.

In summary, technical concerns need to be addressed to fully benefit from reproducibility and replicability features, in addition to the cultural and educational challenges described more fully in Chapter 5.

DISSEMINATION

The scientific article, published in a prescribed format in a peer-reviewed journal, has constituted the essential ingredient of traditional publishing since the 17th century. Although traditional publishing is still dominant, new models are gaining strength, with the opportunity for the publishing community to rethink its approaches, not just tinker and automate existing practices. One important step toward openness was the memorandum on Increasing Access to the Results of Federally Funded Scientific Research from the Office of Science and Technology Policy (Holdren, 2013), which directed federal agencies with over $100 million in annual conduct of research and development expenditures to develop a plan to support increased public access to the results of research funded by the federal

---

[3] For example, CRediT (Contributor Roles Taxonomy) includes 14 contributor roles that are typically played by contributors to scientific scholarly output. See https://casrai.org/credit.

government (NASEM, 2018b). ARWs not only can play a direct role in the processes of how research is disseminated, but also can provide new opportunities (and challenges) in the types of outputs that need effective dissemination and review. They can support the rapid publication of new findings, shown to be critical as the research landscape for COVID-19 barrels forward. The experience and new ways of working can be translated into all fields of research where speed of progress is equally important.

Publishers and other research disseminators have taken important steps to adapt their systems and requirements to foster transparency and access to nonarticle outputs as information technology advances and other shifts have changed research practices. Notable examples include efforts to facilitate data sharing, bring clarity to author contributions, and enable interdisciplinarity and more rapid utilization of research findings through "convergent" approaches such as We Share Data's Data Sharing Seminar Series for Societies[4] (McNutt, 2017; McNutt et al., 2018).

In terms of streamlining publishing workflows, AI-based tools are already being used by most publishers to detect plagiarism (e.g., iThenticate). An increasing number of publishing-related organizations (e.g., Digital Sciences, Clarivate/Publons, Elsevier) have developed services that use algorithms and text mining and/or natural language processing to support authors, editors, funders, and others in identifying relevant peer reviewers, guest editors, contributors to special issues, and others involved in the process. Some AI-based tools aim to accelerate the publication process as well as streamline the effort required to validate article submissions, which can be especially important to make high-priority areas of research available sooner. For example, StatReviewer aims to check that the statistics and methods in manuscripts are sound, and UNSILO's Evaluate tool[5] uses advanced machine intelligence and natural-language understanding to help authors, editors, reviewers, and publishers carry out evaluation and screening of submitted manuscripts.

Many tools use text mining, ML, network analysis, and other methods to filter published research to support researchers in keeping on top of the relevant literature (e.g., Researcher app, most bibliographic reference manager apps). These services identify linked concepts in the literature that may not be obvious to humans and surface emerging trends (e.g., Meta, Euretos). Some reference management tools, such as Sciwheel,[6] support authors by assessing their written text and suggesting other articles to improve the quality and relevant completeness of their citation lists.

ARWs can benefit publication through greater adherence to minimum reporting requirements, and they open the possibility of bringing reporting standards to fields where they are not so advanced. This would reduce time and effort for

---

[4] See https://wesharedata.org/.
[5] See https://unsilo.ai/unsilo-evaluate.
[6] See https://sciwheel.com/?lg.

authors, publication editorial teams, and reviewers in checking whether enough detail has been provided to support reproducibility. *Nature*, for example, has compiled its reporting requirements and resources for contributors, in which the author's responsibilities related to sharing of data, materials, codes, and protocols are emphasized in bold-faced type (*Nature*, 2021).

At the same time, ARWs increase the need for greater agreement on reporting standards so that different workflows and tools capture the same minimum set of information. These requirements also need aligning with the specific depositing requirements of subject-specific repositories that often have minimum metadata requirements; a service called FAIRsharing[7] is one way to monitor this landscape of ever-changing standards (Sansone et al., 2019). This, together with an increasing use of a standard set of ontologies, could help to realize the vision of the European Open Science Cloud and other such clouds of making data truly interoperable between disciplines, and open significant opportunities of truly collaborative research.

A shift in how research is conducted and produced is allowing the community to rethink publishing approaches beyond simply automating existing practices, but rather to better utilize technologies aligned with the way research outputs are being produced. For example, models are now available to rapidly publish (typically within a few days) new findings together with the underlying findable, accessible, interoperable, and reusable (FAIR) data and detailed protocol information, to be versioned and updated as new data come in (e.g., F1000Research).[8] Other platforms include the Gates Foundation's Gates Open Research[9] and Open Research Europe.[10] These models use transparent peer review and support in-article visualization of data and tools such as Code Ocean[11] and Whole Tale[12] so that readers and peer reviewers can assess the code, edit it, and reanalyze the data on the fly within the article without the labor-intensive need to set up the relevant computational environment. Some publishers are also exploring publication of electronic notebooks and including these directly into publishing, and a few in chemistry are including these directly into publishing workflows (AGU, 2021).

The advance of container technology—which "allow[s] packaging up all code and dependencies to ensure that analyses run reliably across a range of operating systems and software versions"—has a lot of promise (Wiebels and Moreau, 2021) in this space. Containers can be disseminated along with articles and enable reproducibility and broader sharing. Workflows can thus effectively be distributed with their entire runtime environment and versioning, allowing for preservation of provenance information.

---

[7] See www.fairsharing.org.
[8] See https://f1000research.com/.
[9] See https://gatesopenresearch.org.
[10] See https://open-research-europe.ec.europa.eu/.
[11] See https://codeocean.com/.
[12] See https://wholetale.org.

COVID-19 has demonstrated the need for such rapid and collaborative research in public health emergencies and the effectiveness and impact of this way of working. For example, in early March 2020 at the request of OSTP, the Allen Institute for AI and partners created CORD-19,[13] an AI-enabled, open, machine-readable collection of papers and data (Wang et al., 2020). As of early July 2020, it held more than 130,000 articles, obviously more than any group of humans could expect to skim, much less truly take advantage of. As another example, to avoid lag time of the normal peer review process but minimize the impact of preprints that would not ultimately withstand such review, MIT Press and University of California Berkeley launched RR:C19,[14] what they characterize as an "open access, rapid-review overlay journal that will accelerate peer review of COVID-19-related research" using AI tools. Such efforts are enabling a transformation of research into the virus and demonstrate how progress can be made in drug and vaccine discovery at speeds many times faster than previously achieved (see additional discussion in Chapter 3). This experience and new ways of working now need to be translated into all fields of research where speed of progress is equally important.

As the speed of research updates created by automated research is likely to increase, we need to consider how to adequately review these outputs, especially given that peer reviewers are already overwhelmed. Publishers are developing article transfer mechanisms of various forms to minimize subsequent review as a manuscript passes between journals looking for acceptance. During COVID-19, new initiatives (e.g., Outbreak Science[15]) have developed to incorporate pre-submission triage mechanisms on preprints to minimize direct peer review requests from already overworked coronavirus experts. But these efforts will not be enough for the potential volume increase from significant uptake of ARWs. Smarter ways are needed to decide what really warrants full peer review as currently practiced versus different types of peer review. These new forms of review may be conducted by different actors in the system (e.g., data curators as reviewers who are experts in automated workflows, rather than just discipline experts), and in some cases could include AI-based peer review. Not only will this widen the pool of potential reviewers, but it will also bring much needed breadth in reviewers' perspectives and expertise as research becomes increasingly multidisciplinary and collaborative.

Developments are emerging in automating the process of creating data note publications (short, peer-reviewed publications that describe research data stored in a repository) in XML alongside rapid data production, which then goes through AI-based peer review to conduct an initial set of checks, followed by community review. A collaboration between Wellcome, the Sanger Institute, and F1000 Research produced the first such publications in 2021. However, in the race to

---

[13] See https://allenai.org/data/cord-19.
[14] See https://rapidreviewscovid19.mitpress.mit.edu/.
[15] See https://outbreakscience.org/.

keep up with the volume of outputs, it is important not to lose the detail of peer review sufficient to adequately assess workflow systems and hence the impact that they could have on research and trust in the scholarly system.

In addition to articles, data, and software, workflows themselves are important research products that can be shared, evaluated, and reused. Challenges related to ensuring FAIR workflows (discussed in Chapter 5) and using AI and ML in ways that are transparent and reproducible (discussed above in Chapter 4) are relevant to developing methods and platforms for disseminating ARWs as expressions of methods underlying reported work. Specific issues include standardizing formats for unstructured data and accommodating AI black boxes—models created from data by an algorithm that are "inherently uninterpretable and complicated" (L'Heureux et al., 2017; Rudin and Radin, 2019).

The use of automated and AI-based workflows also opens interesting questions on the impact on authorship, new types of contributions to the work (e.g., the relationship between the discipline expert and the workflow developers), and the balance in seniority between those roles, as well as credit for work if the automated workflows are generating their own further research questions based on the previous data. With authorship in traditionally published outlets as a significant component toward promotion, tenure, and funding decisions, the balance between the contribution of the workflow and the human needs to be carefully thought through.

5

Overcoming Barriers to Wider Use of Automated Research Workflows

With all the benefits that automated research workflows (ARWs) can provide, whether in discipline contexts as summarized in Chapter 3 or across disciplines as summarized in Chapter 4, the question remains: What inhibits their greater use? Beyond technical challenges, discussion at the March 2020 workshop and other information indicates that the same conditions that slow or prevent change in other aspects of the research enterprise are in play here as well. These conditions include the tendency to maintain academic silos and a focus of research funders on investigator-led projects rather than underlying infrastructure. Perhaps more unique is an overlay of concern about machines taking over from humans—whether that concern is expressed about machines supplanting humans in the quest for discovery or the more prosaic, yet understandable worry about loss of jobs, changes in workplace practices, and obsolescence. As made clear throughout this report, and reiterated below, ARWs are a tool, but people remain central to the process of discovery, no matter the level of automation or terabytes of data.

The committee identified five main challenges to wider use of ARWs and offers ideas to address them related to the incentive system, the current research culture, education and training needs, sustainability, and privacy and ethical concerns.

REIMAGINING INCENTIVES

There has been extensive discussion in recent years about the perverse or misaligned incentives for researchers that result from hypercompetition and the inappropriate use of bibliometric measures in evaluation (Teitelbaum, 2008; Casadevall and Fang, 2012; Stephan, 2012; DORA, 2013; Alberts et al., 2014; NASEM, 2017). Hypercompetition has emerged due to slower growth in research funding alongside continued high production of Ph.D. degrees in science and engineering, leading to a scarcity of tenure-track positions in academic research relative to the number of qualified candidates. Some experts believe that the use of bibliometric measures such as the Journal Impact Factor in evaluation leads students and early career researchers to focus their efforts on publishing articles in the most prestigious journals and to choose currently fashionable topics where articles are likely to be highly cited, rather than to engage in riskier fundamental studies (Lawrence, 2007; Alberts, 2013; Ioannidis et al., 2014).

As a result of this emphasis on publishing in the most prestigious journals in the reward structures of many researchers, the current research ecosystem tends to reward researchers for independence and innovation over practices that support rigor and reliability. Publishers themselves have made significant efforts to encourage data sharing over the past decade and have established community norms for rigor and transparency in data generation (see discussion in Chapter 4). However, these expectations are not yet applied to other components of the research life cycle. For this reason, independent and artisanal experimental procedures are typically used in research and often lack the oversight necessary to truly evaluate the validity of the experimental methodologies and/or implementations used.

The integration of ARWs into the research life cycle provides an opportunity to address these issues. However, doing so will require a shift in incentives for scientific productivity. Research culture and values are set by funders, academic institutions, publishers, regulators, and tooling platforms, and are often affected by national policies. These entities will play a major role in changing researcher incentives in ways that foster more rapid and effective adoption of ARWs and ensure that community standards for quality assurance, transparency, and reproducibility are upheld.

Incentive structures were mentioned as barriers to progress across most of the use cases covered in Chapter 3. For example, many fields do not have a strong tradition of sharing data, and the task of curating data for others to use is not generally rewarded and can therefore prove hard to justify by time-strapped researchers. This hinders progress in building the large, artificial intelligence (AI)-ready data resources needed for ARWs in most fields. Senior researchers may not believe that AI always provides meaningful intuition into foundations or reasons of association and may discourage junior researchers from pursuing it. AI that is oriented to provide causal reasoning may address this issue.

Examples of the sorts of shifts in incentives that would create a positive environment for implementing ARWs include

- Weighing reliability and innovation equally when evaluating research productivity.
- Rewarding the sharing of a much wider range of outputs beyond standard narrative research articles (e.g., data notes, software tools, and methods articles) as well as valuing negative or null findings to ensure a balanced

representation of research understanding and generation of knowledge from those findings.
- Valuing transparency and reproducibility through all phases of the research life cycle.
- Rewarding the use of validated and documented experimental processes.
- Encouraging researchers to incorporate process design and systems analysis into their experimental workflows to ensure that all components of the research life cycle are performed and disseminated in a transparent and reproducible manner.
- Rewarding researchers for the measurable reuse of workflow system-enabled results for applications that employ algorithms and workflows.
- Incentivizing collaboration and team science.

OVERCOMING BARRIERS IN THE RESEARCH CULTURE

Cultural changes in the research enterprise are necessary for effective adoption and use of ARWs. It will be important to develop these processes in a way that promotes ARWs as tools that can support both reliability and innovation in discovery, rather than falling into the trope of "machines replacing humans." This inaccurate representation has been seen extensively with the advent of AI in medicine, including articles in the popular press about whether "AI will replace doctors," and it has hampered progress.

Current research culture focuses on the scientific laboratory and the principal investigator as an independent artisan responsible for inventing innovative solutions to all aspects of a problem. In fact, the approach leads to unnecessary uncertainty in experimental steps. This is particularly true for processes that can easily be standardized and automated. ARWs are reliant on the stitching together of individual components, and, by necessity, these each need to be as performant as possible. The explosion of fully annotated research data that will accompany the emergence of ARWs will facilitate collaborations between researchers and across laboratories, teams, or departments. One can easily imagine a future network of connected ARWs that use distributed AI and consensus learning to mutually coordinate massive experiments that would have been previously inconceivable or impractical.

Training researchers to view themselves as the master regulator of the entire system should ensure that they innovate where innovation is necessary and useful—rather than innovating artisanal solutions to solved problems that should be standardized. A combination of shifts in top-down expectations (policy, funding, and training) and bottom-up expectations (build it into the tools, directories for finding things, peer review expectations, and society-level shifts in focus of innovation) should facilitate this culture shift. As noted above in the discussion of incentives, research funders can support this culture shift by emphasizing (and rewarding) transparency, reliability, and teamwork in their granting practices.

One central task for researchers as managers of ARWs will be to understand the interplay between reductionist science and its parameter-sparse models, which have been immensely successful, and, on the other hand, deep learning methods, whose success relies on overparameterization. What we need is the best of both worlds: machine learning (ML)-informed science. As discussed in Chapter 2, this requires continued methodological advances. We need to find good ways of imposing scientific structure on deep learning models, or, conversely, inform scientific models through ML.

Finally, as noted several times throughout this report, it will remain important to leave space for serendipity. Automated workflows, community standards, and collaborative approaches are tools designed to support researchers in reliable scientific innovation. They should be used where appropriate but should not replace active and individual human oversight where needed to detect the unexpected.

CLOSING EDUCATION GAPS

Future generations of ARWs will have a significant impact on research practice and scientific experimentation, analysis, and interpretation. However, to pull maximum benefit from these systems, future researchers will need additional training outside their discipline. This will require integrating domain science training with data science training and relevant software engineering into academic programs across all disciplines at both the undergraduate and graduate levels. In addition, research teams will need additional specialized expertise from research software engineers, computational scientists, and data stewards.

The use cases discussed in Chapter 3 revealed some education and training needs and challenges common across multiple domains. A need for more researchers who combine domain knowledge and data science or software development expertise was expressed in just about all the use-case discussions. In several of the use cases, including materials research and chemical synthesis of pharmaceutical compounds, presenters expressed a growing need for researchers who are knowledgeable in working with automated laboratory equipment. For example, Rebecca Nugent described Carnegie Mellon's recently established graduate degree program in automated science, and Carole Goble noted the long-term return on investment in establishing a "career track for research software engineers." More broadly, knowledge of mathematical and computational methods for designing experiments or controlled observations automatically and for learning from data is becoming essential for researchers. Several universities (e.g., Massachusetts Institute of Technology and California Institute of Technology) are beginning to address this need by establishing cross-disciplinary graduate programs or cross-links between domain-specific programs and programs in the computational and mathematical sciences. By contrast, there will be less need for researchers to perform mechanical activities that have traditionally been a part of standard training—such as pipetting in biology—and more need to understand the overall goals of the research program and design a complex approach to reaching the goals using more automated tools and technologies.

The use cases discussed in Chapter 3 also illustrate the need for additional specialized expertise in data stewardship, software development, and other computational tasks. For example, research software engineers are key players in the development of ARWs and other research workflows, with their own career paths. Organizations such as the United States Research Software Engineer Association,[1] the Society of Research Software Engineering,[2] and the Campus Research Computing Consortium[3] are working to build community among research computing and data professionals. A report from the Organisation for Economic Co-operation and Development emphasizes the critical importance of data-intensive science and the need to strengthen the digital capacity and skills of the scientific enterprise (OECD, 2020).

Students in different disciplines may require different types of supplementary data science training and at differing depths. However, it is likely that all students will benefit from increased exposure to workflow-related data science areas including, as described at the committee's workshop, statistical modeling of high-dimensional data (Nugent), scientific ML ("AI 3.0," Vidal), database management (Wietzner, 2020), sensor management (Beckman, 2020), FAIR workflows (Goble et al., 2020), and the application of AI for hypothesis generation and automation (Stodden, 2020). Advanced statistics training will enable researchers to cogently interpret the strength of evidence for the complex high-throughput experiments that are run by ARW systems. Training in managing and analyzing data, including areas such as how to model data and select the right tools to support particular models, will be required for managing and accessing the massive and diverse data repositories that workflow-aided science is expected to produce. Training in AI will support researchers in evaluating and selecting the appropriate black-box ML models used by automated workflows for extracting knowledge and designing new experiments and studies.

Many disciplines have begun the process of introducing data science as a core component of academic training, for example, experimental chemistry, astronomy, and biology (Cernak, Glotzer). Training in ARWs will be a natural extension of this process in these disciplines. The transformation of science education will foster new collaborations with faculty in computer science, statistics, and mathematics and will result in the creation of new courses that integrate data science and experimental science (Stodden, 2020). Faculty will need to grapple with difficult questions about existing domain science topics that will need to be dropped to provide time to introduce workflow topics. Yet doing so will equip the

---

[1] See https://us-rse.org/.
[2] See https://society-rse.org/.
[3] See https://carcc.org/.

next generation of students to assess the tradeoffs between using physical models versus AI prototypes in designing scientific workflows, and to weigh the sacrifice in model interpretability against the convenience of black-box automation of experiments (Vidal). Although it is not necessary for all discipline experts to acquire expert proficiency in data science or coding, they should have enough background to critically assess the "black box" aspects of many workflow tools so they can understand and make adjustments for any likely biases inherent in the system.

One of the March 2020 workshop speakers suggested a core curriculum for research and development skills relevant to ARWs leading to a certification that would include process design, measurement systems analysis, process qualification, design of experiments, and quality by design (Gardner, 2020). Such an approach would provide grounding in both the principles underlying ARWs as well as implementation issues. The curriculum could include both self-learning and classroom learning and include variants for different educational levels. Other perspectives that might be incorporated into new ARW-friendly curricula include the potential of ARWs in translational research and convergence research. The concept of translational research—meaning the conversion of basic knowledge into products or processes that meet critical real-world needs—emerged several decades ago in the biomedical domain. Computational scientists have begun to conceive of priorities within their own discipline, including workflow management systems, as examples of translational research (Deelman et al., 2020). According to the National Science Foundation (NSF), convergence research is characterized by integrated work across disciplines directed at a specific, compelling problem (NSF, 2020a).

However, as noted above, changes in institutional culture and incentives will be needed to accelerate and sustain such transformations in education. At many research universities, existing tenure and promotion criteria do not favor faculty collaborations on programmatic or curricular transformations. Department chairs are likely to be concerned about letting their faculty divert their efforts away from departmental teaching commitments. Universities may lack the resources to support the major investments needed to revise domain science curricula to include advanced workflow training. These hurdles can be overcome with extramural financial support from foundations, federal agencies, and industry (Kusnezov, 2020). National Institutes of Health (NIH)-like training grant programs would be an enabler of faculty and student investment into workflows for research (Cernak, 2020). Partnerships between academic institutions and industry will be especially fruitful since industry has led the use of workflow technology in scientific research and development (Fox, 2020).

ENSURING SUSTAINABILITY

A common concern voiced at the workshop focuses on how to fund ARW development and maintenance to support research within and across domains. Realizing the promise of ARWs requires investment in hardware, software, and human resources. Although there seems to be general agreement that such investment is important, the current level and structure of resource allocation fall short of these well-stated intentions in the realm of real-world funding.

Investment Priorities to Advance ARWs

For several of the use cases discussed in Chapter 3, the development of tools and technologies constitutes a key enabler for accelerating progress. For example, materials researchers examined existing research workflow management systems and ended up building their own due to the need for a system that enables dynamic rerouting, facilitates constant communication among researchers, incorporates error management capability, and is flexible. Building an advanced computational environment for wildfire monitoring and behavior prediction requires integrating numerous functions, such as collecting various types of data and performing multiple, complex modeling tasks. In particle physics, development of computational tools that allowed for collaborative statistical modeling, in addition to workflow management and computational tools, was critical to confirming the existence of the Higgs boson. Laboratory automation technology is a major driver for advances in experimental domains such as chemical synthesis of pharmaceuticals and materials research.

To realize the potential of ARWs, it is essential that software tools for aspects of workflows that transcend disciplines—for example, those involving AI and ML methods for designing experiments and learning from data—become interoperable and broad purpose. This will require a level of software engineering not commonly encountered in academic environments. Usually, research groups develop software for their own purposes, but have neither the means nor the incentives to build scalable software platforms that can reach beyond an individual laboratory. Universities generally lack the software engineering infrastructure to develop and maintain the complex, interoperable, and performance-portable code bases needed to realize the full potential of ARWs. A sustainable infrastructure requires software engineers, test engineers, and release engineers, but none of these are typically funded in research grants. Several institutions are working on software sustainability, such as the Software Sustainability Institute, WSSSPE (Working Towards Sustainable Software for Science Practices and Experience), and the Workshop on Sustainable Software Sustainability, and the general problem is that software stops working if not actively maintained (Hinsen, 2019).

The availability and utility of data and related infrastructure such as repositories and active curated services are also critical to the implementation of ARWs in many fields. In both experimental and observational fields, including materials research and astronomy, FAIR data are needed to develop and train ML algorithms, which in turn enable the development of closed-loop systems in which the selection of experiments or instrument targeting can be automated. In some fields, such as particle physics, there is considerable experience with collecting and processing large amounts of data, but new approaches to instrument design and data are needed to allow for simulation-based inference based on reuse of data. Across several of the fields examined, including digital humanities, there is a growing need for shared community data resources, such as FAIR repositories. One of the workshop speakers cited digital music as an analogy; to implement ARWs, communities need to move to shared data resources in the cloud that are available for a myriad of uses, similar to music streaming services. Creating and sustaining community data resources involves many challenges, including funding, deciding which data sets should be stored and maintained, and facilitating interoperability between them. The lack of availability of these resources ultimately limits the size and scope of collaborations.

Realizing FAIR Data, Software, and Workflows

Another aspect of fostering sustainability is support for community efforts to develop new tools, services, and frameworks aimed at realizing FAIR data, software, and workflows. As seen in the use cases discussed in Chapter 3, data that are FAIR, well curated, discoverable, and actionable do not just appear. Significant directed investment and specific actions are needed to support creation, sharing, and curation of such data at volumes needed for implementation of ARWs, and use of ML and AI as research tools at a significant scale in many domains. Actions on the part of several stakeholder groups could help address this.

To begin, publishers can move away from the acceptance of data supplements as adequate for fulfilling their data-sharing requirements and instead directly associate articles to data in FAIR repositories, with preference given to leading domain repositories. One initiative in the earth and space sciences is working to implement this approach, which could be adopted by other disciplines (Stall et al., 2019; COPDESS, 2021).

In addition, funders could increase support for leading domain repositories, for the creation of new repositories, and for the broader data ecosystem. Leading domain repositories provide quality FAIR curation and simplify discoverability. They also help develop leading practices around what data should and can be preserved. Quality curation and metadata will enable interoperability. Many domain repositories are poorly or inconsistently funded and thus are forced to spend significant staff time on fundraising that could be spent on data services. Support is also needed for related organizations that provide important infrastructure for the data ecosystem, such as Crossref, Datacite, the Research Data Alliance (RDA), the National Information Standards Organization, and FORCE11.

Research communities can work to ensure that domain and institutional repositories collaborate effectively. There are potential mutual benefits to be gained when institutions support leading repositories as illustrated by the California Digital Library's (part of the University of California) partnership with the Dryad Digital Repository (Waibel, 2018). Such collaboration will need to be supported.

Funders and institutions can also do more to incentivize researchers to practice FAIR principles for data and other research outputs such as software (FORCE11 is approving software citation guidance and much direction exists). This can be done through the grant process and data management plans (DMPs) and broader efforts to communicate the value and importance of FAIR principles as part of changing the institutional culture. One promising example is Duke University's effort to provide various services and training opportunities to researchers aimed at improving the skills of researchers in developing and implementing DMPs (Duke University, 2021).

Funder requirements around data and DMPs can also evolve over time so that fulfilling data-sharing and management expectations are recognized and appreciated as key outcomes of the award. Examples of language that funders can use to encourage sharing of data and other research products such as software, methods, and samples were discussed as part of a recent National Academies workshop and resulting proceedings on Developing a Toolkit for Fostering Open Science Practices (NASEM, 2021). Additional community thinking on these issues is contained in the responses to the White House Office for Science and Technology Policy Request for Information on Draft Desirable Characteristics of Repositories for Managing and Sharing Data Resulting from Federally Funded Research (OSTP, 2020).

Another set of tasks relates to realizing FAIR principles for workflows themselves. For example, there has been considerable progress in community efforts to develop standards in areas such as registries (Dockstore, an app store for bioinformatics;[4] WorkflowHub,[5] a registry for describing, sharing, and publishing scientific computational workflows), services for monitoring and testing (LifeMonitor,[6] OpenEBench[7]), standards for packaging (Workflow-RO-Crate[8]), and Bioschemas' schema.org definitions for workflows[9] (Goble et al., 2020).

Efforts to develop standards for describing workflows, such as Common Workflow Language (CWL) or Workflow Description Language (WDL), and more general abstractions for recovering workflow information from scripts are also important for achieving FAIR workflows (McPhillips et al., 2015; Perkel, 2019). The Global Alliance for Genomics and Health[10] is creating standards for defining, sharing, and executing portable workflows. All of these services are implemented in the European EOSC-Life Workflow Collaboratory,[11] for example, and similar services exist elsewhere. These are the necessary components of

---

[4] See https://dockstore.org/.
[5] See https://workflowhub.eu/.
[6] See https://crs4.github.io/life_monitor/.
[7] See: https://openebench.bsc.es/dashboard.
[8] See https://about.workflowhub.eu/Workflow-RO-Crate/.
[9] See https://bioschemas.org/.
[10] See https://www.ga4gh.org/.
[11] See https://www.eosc-life.eu/.

FAIR workflows. Another relevant standard is the IEEE 2791-2020 Standard for Bioinformatics Analyses, a regulatory metadata framework for the U.S. Food and Drug Administration's High Throughput Sequencing workflows for precision medicine (IEEE, 2020).

Why Is It Difficult to Secure Sustained Support for These Priorities?

One strong theme of the March 2020 workshop discussion is the difficulty in securing sustained support for the sorts of efforts described above. Even in the case of successful enabling tools such as Jupyter, the funding situation is sometimes tenuous (Granger, 2020). Since World War II, the primary modes of federal research funding in the United States have included short-term competitive awards to individual investigators organized by discipline by agencies such as NSF and NIH, longer-term funding of large intramural and extramural projects that advance the missions of agencies such as the Department of Energy (DOE) and the National Aeronautics and Space Administration, and approaches aimed at catalyzing progress toward addressing particular issues relevant to an agency mission (e.g., Defense Advanced Research Projects Agency [DARPA], IARPA, ARPA-E).

Given this framework, providing sustained support for the development and operation of shared infrastructure that serves multiple disciplines has historically been challenging. Nevertheless, since the launch of the Advanced Research Projects Agency Network (ARPANET) in the 1960s, there have been several examples of the U.S. federal government providing shared resources that have allowed research communities to harness information technologies to significantly advance their work. Examples include the establishment of national supercomputer centers in the 1980s by NSF in partnership with academic institutions, the development of GenBank and other digital data resources in the life sciences by NIH and the National Library of Medicine starting in the 1980s, and NSF's advanced cyberinfrastructure program launched in the early 2000s. Current programs aimed at bridging the gaps between investigator-focused projects include Harnessing the Data Revolution (NSF) and Big Data to Knowledge (NIH), and a series of efforts to advance strategic computing and related technologies across agencies under the auspices of the Networking and Information Technology Research and Development Program and its predecessors. Relevant programs and efforts by DOE, DARPA, and international efforts on the part of the European Union and UK Research and Innovation were also discussed at the workshop and are highlighted in Chapter 2.

It is difficult for individual institutions to make cyberinfrastructure investments in the same way as they view, for example, a mass spectrometer or other large physical "thing." One possible model is the Harvard Dataverse, with tens of thousands of data sets deposited for sharing and over 1.5 million downloads. Internet2 is an example of a nonprofit university consortium that has provided sustained support for information technology infrastructure under a membership model.

Note that the above barriers to sustainable investments are not only relevant to ARWs but apply to all research that depends on software, which is a significant fraction of all research (Nangia and Katz, 2017). As discussed in Chapter 2, it is inherently more difficult to fund development and maintenance of production-quality software (workflow engines, automated tools, etc.) that can be used broadly than to develop new software as part of a research project that may not be used outside that project.

The commercial and nonprofit sectors support ARWs in various ways. The challenge is finding a route that provides continued access and availability to researchers while recognizing that these backers have their own revenue- or mission-based goals to meet. Big tech companies have provided cloud computing services at a large scale to academic researchers at low cost. There is a potential catch-22 in this support, however, in that at some future point, they can raise costs or restrict use through patents. The workshop also revealed areas where the dominance of proprietary tools or data acts as a barrier to ARWs. One example from materials science is scanning transmission electron microscopy. Current commercial electron microscopes "tend to down-sample or discard the vast majority of the signal via averaging or decimation," and important sample properties are lost in the process (Somnath et al., 2019). Further, the "down-sampled data are usually written into proprietary file formats, which impede and sometimes even preclude access to data and metadata, complicate long-term archiving, obstruct sharing, and fracture the scientific communities along file formats" (Somnath et al., 2019).

Future Directions for Sustainability

It is beyond the scope of this committee to propose specific funding mechanisms or amounts, but rather to emphasize the importance of support at all stages of the pipeline. Indeed, there are ways forward to ensure sustainability. Several speakers at the workshop suggested that providing continued support for ARWs could be a pivotal role for the government. As examples in Europe and the United States, two large consortia have sustained their systems through collaborations and shared resources: CERN, the European research program in advanced physics, and CERT, the Community Emergency Response Team in the United States. Another suggestion was to create a stable endowment, akin to the Smithsonian Institution, that can serve as a common resource.

Given the significant investments that governments and other research funders are making in data-driven science, it makes sense to leverage these investments across borders and domains to the extent possible. Goals of enhanced international collaboration would include facilitating access to tools and resources and ensuring the interoperability of national implementations. Distributed international efforts working to develop standards and approaches to facilitate FAIR data and software include GO FAIR, the RDA, and the Research Data Framework (NIST, 2021). GO FAIR was established to advance the FAIR principles, which emphasize the importance of machine readability and reuse of data (Wilkinson et al., 2016). RDA was started in 2013 and aims to build "the social and technical infrastructure to enable open sharing and re-use of data."[12] The Research Data Framework[13] was initiated by the National Institute of Standards and Technology in 2019 and is aimed at increasing the supply of trustworthy research data across domains by developing a "strategy for various roles in the research data management ecosystem." Additionally, the FAIR for Research Software Working Group is convened as an RDA Working Group, FORCE11 Working Group, and Research Software Alliance Task Force.

MANAGING ISSUES OUTSIDE OF RESEARCH THAT AFFECT ARWS

The March 2020 workshop featured discussion of several important issues outside of the research enterprise that bear on the development and implementation of ARWs. The most obvious issue is the treatment and use of data that are collected from or about individuals. Discussion about the use of ARWs also connects to broader considerations about the use of AI in various societal and decision-making settings.

ARWs promise to open significant new areas of research through the use of data that have not been collected through experiments or simulations, but rather concern the health and medical condition of individuals, their social media behaviors, financial transactions, and the like. In response to concerns about the security and use of personal data—exacerbated by well-publicized examples such as the data breach at credit reporting firm Equifax in 2017 and the 2018 exposure of the use of Facebook data for political purposes by Cambridge Analytica— policy makers and public interest groups have pushed to allow individuals to have greater control over the use, storage, and reuse of their data.

The European Union's General Data Protection Regulations, put into effect in 2018, are intended to protect personal data by placing strong regulations on the entities that collect, process, and use data (EU, 2018). The California Consumer Privacy Act, which went into effect in 2020, established privacy rights that businesses operating in California or that provide a product or service to a resident of California must take steps to protect (Rothstein and Tovino, 2019). Although this does not cover research in the public interest, as of this writing, there was still some ambiguity about how the law will affect the research community, with some clarifying legislation proposed (Moundas and Peloquin, 2020).

---

[12] See https://www.rd-alliance.org/about-rda.
[13] See https://www.nist.gov/programs-projects/research-data-framework-rdaf#:~:text=To%20address% 20these%20issues%2C%20NIST%20initiated%20a%20new%2C,customizable%20strategy%20for%20 the%20management%20of%20research%20data.

At the same time as a privacy-aware public shares personal information in unprecedented ways through social media and other avenues, many also express reservations about its use in research. Institutions and commercial entities have established their own policies and requirements about data use for research. As one workshop participant pointed out, academia has tended to be more cautious than the private sector about the use of some personal data, such as social media data, in research (Weitzner, 2020).

Some common work-arounds, such as de-identification, can impinge on research, most notably biomedical research. As Bradley Malin, director of the Health Information Privacy Laboratory at Vanderbilt University (and member of this committee) noted at a 2018 workshop on planning for long-term use of biomedical data, "De-identification results in a loss of data utility; encryption results in a loss of functionality; and secure environments result in a loss of efficiency. However, with no action, the potential outcomes include losses of privacy, money (due to litigation and remuneration), societal trust, and scientific opportunity" (NASEM, 2018a). An irony is, as one workshop participant noted, "success can lead to failure." By this he meant that workflows may be able to use data that were not previously being used, whether because an experiment did not have the intended results or simply because of the data's size, unwieldiness, or seeming irrelevance. As the data take on a new purpose (and value), the data owner may take another look and place restrictions on what was heretofore more accessible.

Even areas of research that do not directly work with personal data must consider privacy issues. The goal of making the workflow itself transparent strengthens reproducibility but could impinge on privacy under certain circumstances, for example, by revealing personal information about specific researchers. As one workshop presenter pointed out, tracking and crediting provenance in data generation must address these types of privacy issues (Weitzner, 2020).

Emerging blockchain-based approaches to using and analyzing data hold the potential to alleviate some of these privacy concerns. For example, federated learning is "a new framework for Artificial Intelligence (AI) model development that is distributed over millions of mobile devices [providing] highly personalized models" while protecting privacy (Bhattacharya, 2019). A collaboration of European companies and academic research institutions is developing MELLODDY (Machine Learning Ledger Orchestration for Drug Discovery), an ML platform using federated learning to allow participating organizations to use proprietary data to speed drug discovery while data owners retain control of those data (IMI, 2021).

The development and implementation of ARWs also involves broader issues raised by the growing use of AI and ML in a variety of policy-making and decision-making contexts. How can transparency and trust in outcomes with significant real-life consequences be maintained when the characteristics of a specific algorithm unrelated to the issue or problem determine those outcomes? Stoyanovich et al. (2020a) suggest the need for a framework to connect interoperability and trust in algorithmic-based decisions.

Initiatives have been launched in recent years under the rubric of "responsible AI" and "responsible ML." For example, Fairness, Accountability, and Transparency in Machine Learning (FAT/ML) is a series of workshops aimed at exploring the challenges raised by ML "for ensuring non-discrimination, due process, and understandability in decisionmaking."[14] Organizations such as Google have developed principles for responsible AI, and the Institute for Ethical AI & Machine Learning was established in the United Kingdom to carry out "highly-technical research into processes and frameworks that support the responsible development, deployment and operation of machine learning systems" (Google, 2021; Institute for Ethical AI & Machine Learning, 2021). The principles and guidelines espoused by these initiatives overlap to a significant degree, with the need for human review, protecting data privacy and security, uncovering and addressing bias, and support for transparency and reproducibility generally being invoked.

The technical, legal, and policy barriers to implementing ARWs are intertwined in the sense that technological development needs to be informed by policy and legal requirements. Possible approaches to addressing these issues are both computational and policy or legal related. New laws or regulations could bring clarity and uniformity to software services as providers work to comply. This would constitute a type of convergence or use-inspired research. An example discussed at the workshop was the changes in computer hardware and software related to accessibility for people with disabilities. The Americans with Disabilities Act spurred the kinds of technological change that are now integrated in virtually all computer systems and tools.

It was also suggested that governments can play a role as an honest broker for data use. Here, too, however, some tension was recognized between agencies that mainly fund and undertake research that produces presumptively open data and agencies that are oriented toward work that produces more restricted data (some agencies do both). Research and its associated data production and use or reuse is also international, making the effectiveness of a single national government on shaping global policy challenging. Data use agreements (DUAs) can address many of these issues from the start, rather than as an add-in consideration, especially when partnerships are formed across the public and private sectors (O'Hara, 2020). However, DUAs are often difficult and time-consuming to conclude (Mello et al., 2020). Privacy, ethics, and similar socially based topics will likely emerge as more data are added into the workflow cycle, and the questions being asked are modified. Building mechanisms into ARWs that recognize and are sensitive to data use issues is new territory to explore and develop, for example, to distinguish between permitted and prohibited queries (Kusnezov, 2020). Ideas proposed at the workshop included embedding compliance in the design of the software for open research data services and standards for the architecture of the sharing and access system (Burgelman, 2020).

---

[14] See https://www.fatml.org/.

In addition to privacy concerns, attention has focused on algorithms that perpetuate sexism, racism, and other forms of discrimination and hate. Well-known examples include Microsoft's TayTweets, which "learned" so much vitriol that it had to be shut off within 16 hours (Hunt, 2016), and the perpetuation of biases in ML (Caliskan et al., 2017). As Oren Etzioni, the CEO of the Allen Institute for Artificial Intelligence, told attendees at a National Academies–convened workshop in 2018, "systems use data from the past to generate models to predict the future, so if society's past was racist and sexist, the models will carry that bias into the future and also, for technical reasons, exacerbate it" (NASEM, 2018c). Preventing it, he noted, requires human attention and intervention.

In the humanities, Hepworth and Church (2014) looked at two data visualizations of instances of lynchings and other white supremacy mob violence that depicted two different results about the extent of these acts of terror, particularly when looking at history of the U.S. West, based on decisions made by the researchers setting the parameters of the data searches. They used this example to propose an "ethical visualization workflow" with three main phases: pre-data collection (defining, reviewing); data collection and curation (collecting, pruning, and describing); and data visualization and argumentation (surveying, previsualizing, visualizing, and publishing). The workflow can be implemented, the authors argue, with a twofold approach that is similar to other areas in which domain and computer specialists must work together:

> firstly, by familiarizing themselves with the latest research in the content field and adjacent field; secondly, by including team members familiar with the entirety of the data pipeline from collection to cleaning to presentation . . . User experience design is particularly important for evaluating the interpretive intervention made by the visualization and mitigating harm caused by the final visualization. (Hepworth and Church, 2014, para. 43)

In lab-based science, an intervention to embed research ethics training was evaluated through a randomized trial conducted by the Center for Open Science (COS) and the University of California, Riverside, with a grant from NSF (Plemmons et al., 2020). In the intervention, training in responsible conduct of research was integrated into the ongoing projects and circumstances of the laboratory, rather than an isolated training in a classroom or online. The training, called the Institutional Re-Engineering of Ethical Discourse in STEM (iREDS), was developed by COS and is available free and online. While it goes beyond workflow development and use, the two areas of the training that were the focus of an article by Plemmons et al. (2020) are relevant: author attribution and data management. The authors concluded, "The iREDS approach shifts the paradigm of research ethics training from merely telling researchers what is and is not ethical, to empowering them to incorporate ethical practices into their research workflow."

6

Conclusion

The tools and techniques being developed under the large umbrella of automated research workflows (ARWs) promise to collapse the centuries-old serial method of research investigation into processes where thousands or even millions of simulations or experiments are iterated rapidly in closed loops, with the analysis of data and even the design of experiments or controlled observations being assisted by machine learning (ML) or optimization techniques. Simultaneously, ARWs provide a way to satisfy pressing demands across fields to increase interoperability, reproducibility, replicability, and trustworthiness by better tracking results, recording data, establishing provenance, and creating more consistent metadata than even the most dedicated researchers can provide themselves. The committee's exploration of ARWs illustrates that the research enterprise stands at an important inflection point. The scientific revolution of the 17th century ushered in an unprecedented era of human progress, leading directly to discoveries and innovations that have transformed tasks requiring the application of muscle or simple technologies into services performed by ever more effective machines. The research enterprise will need to develop new approaches and tools as it enters an era in which core elements of knowledge discovery itself can be automated and accelerated.

In important ways, this emerging process of innovation and adaptation represents a continuation of the long-standing trend of computational power being harnessed to perform a variety of research tasks. Yet new twists will need to be considered and addressed. Concerns about privacy, ethics, and trust arising in many domains of human activity become even more relevant to the entire research enterprise as we increase use of artificial intelligence (AI)-based technologies.

As illustrated in the use cases examined in Chapter 3, different disciplines of research have very different usage patterns relative to ARWs—in terms of specific tools and platforms and, more generally, propensity to incorporate workflows into their processes in the first place. Costs for equipment, software, staffing, and training may vary by discipline, but the broad need for domain researchers to incorporate new methods and approaches holds across the use cases. In addition, additional specialized expertise in areas such as software engineering, algorithm development, and data science will be required in a number of fields.

Further, several lines of thought that emerged from the March 2020 workshop are germane not just to the task at hand, but more broadly across the scientific enterprise. These themes include the need to break down academic silos, provide incentives for greater collaboration among researchers, ensure greater interoperability across technologies, foster sharing of a broader range of research outputs, and address issues such as striking an appropriate balance between access to and protection of data.

The committee's findings and recommendations point to promising areas of focus for the research enterprise in facilitating the effective implementation of ARWs. The use cases and supporting literature described in Chapter 3 support all of the recommendations, with Findings A and B and Recommendation 1 in particular flowing directly from examples drawn from a variety of domains. Finding C and Recommendations 2, 3, and 4 are supported in Chapters 4 and 5, which draw on presentations from the March 2020 workshop and other cited literature. Finding C and Recommendation 5 are also supported mainly in Chapter 5, again, with points drawn from the use cases.

FINDINGS AND RECOMMENDATIONS

Finding A: Accelerating Discovery

In many disciplines, the emergence of automated research workflows (ARWs), built upon contemporary cyberinfrastructure, is demonstrating the potential to vastly increase the speed and efficiency of a range of research activities. These include designing and conducting experiments, analyzing data, and observing natural phenomena. These improvements can be realized at scale by implementing infrastructure and practices that facilitate the application of artificial intelligence and machine learning and related technologies to research. Realizing the potential of ARWs could accelerate the pace of scientific discovery by orders of magnitude and thereby expand the research enterprise's contribution to society.

Finding B: Additional Benefits

In addition to increasing the speed and efficiency of research, the effective development and implementation of the technical and human infrastructure for automated research workflows (ARWs) will contribute to strengthening the research process in other ways. For example, the greater transparency and repeatability made possible by automating and capturing specific steps in the research process—advances that underlie the development of ARWs—can foster reproducibility, replicability, and responsibility in research. Adoption of common and interoperable tools and platforms—which could be accelerated by the advance of ARWs but depends on other developments as well—can facilitate international and interdisciplinary research collaboration. Broader access to research workflows and results and the enhanced ability to uncover and correct errors can contribute to greater confidence in research findings and the research enterprise and reduce redundancy among research efforts. To be sure, issues such as dealing with large amounts of streaming data and complex computational approaches will continue to pose technical challenges to the design and implementation of ARWs. In addition, incorporating emerging principles and guidelines for responsible artificial intelligence and machine learning advocated by various organizations, such as building in human review of algorithms, uncovering and addressing bias, and supporting transparency and reproducibility, will also help to secure the benefits of ARWs.

RECOMMENDATION 1: Design Principles

Organizations that fund, perform, and disseminate research, along with scientific societies, should support and enable automated research workflows (ARWs) that embody the following design principles:

- ARWs and the systems, tools, and platforms that comprise them should facilitate openness, reproducibility, and transparency.
- ARWs should facilitate the effective use of artificial intelligence (AI) and machine learning (ML) as research tools and incorporate principles of responsible AI and ML to mitigate the risks from various human and technological deficiencies, such as confirmation and sampling biases, inappropriate application of statistics, and challenges to interpretability of results and quantification of confidence and uncertainties when drawing inferences from ML analyses.
- The associated research objects (data, code, even entire workflows) for ARWs should be FAIR (findable, accessible, interoperable, and reusable), not only by humans but also by machines, to facilitate automated reuse and collaboration.
- ARWs should prioritize reuse and sustainability of existing tools and systems when possible and appropriate, reducing costly duplication efforts and facilitating the extension of capabilities through integration or federation of systems, and agreement on standards. Designs should allow for specialization into specific domains, but avoid unnecessary rebuilding.
- While proprietary services and components can enhance the utility of ARWs, key ARW infrastructure should be controlled by and be accessible to the research community itself, with the community developing standards and practices to facilitate this.

Finding C: Research Enterprise

Realizing the potential of automated research workflows (ARWs) will require modification of the research enterprise, including sustainable funding for the necessary hardware, software, and human resources, educating the scientific workforce, reporting and sharing research results, and structuring researcher rewards and incentives. Multidisciplinary, multirole collaboration is essential to realize the potential of ARWs.

RECOMMENDATION 2: Infrastructure, Code, and Data Sustainability

Research funders, working with other stakeholders such as societies, research institutions, and publishers, should place greater priority on approaches to ensuring the creation and sustainability of key systems, tools, platforms, and data archives for automated research workflows (ARWs). Priorities include

- Funding support for efforts by research institutions and societies to link disciplines so they can share and benefit from the expertise in statistics, machine learning, or data science, and engineering and computer science that is required to build and maintain sustainable infrastructure for ARWs.
- Funders and research communities structuring funding for cyberinfrastructure projects such as large scientific instruments so as to maximize the potential for innovation in ARWs and the reuse of data and other outputs.
- Funders and research institutions supporting open data standards and open interfaces for scientific instruments.
- Funders and research institutions enabling reuse, reproducibility, and long-term sharing of FAIR data and software resources through support of repositories that make archival and updated versions of these resources available within and across disciplines, and providing approaches to sustain those repositories.
- Publishers updating their data-sharing requirements by directly associating articles to data in FAIR repositories.

RECOMMENDATION 3: Human Resources

Research funders, higher education, research institutions, and scientific and professional societies should support the development and implementation of educational programs and career pathways aimed at building the workforce needed to develop and utilize automated research workflows (ARWs), including the creation of career tracks that support ARW capabilities. Examples of what is needed include

- Programs that foster integration of domain expertise with data science and software engineering skills.
- Programs that inculcate data literacy and computational analytical skills in all areas of research.

- Developing the human resources needed to build, maintain, and operate ARW hardware and software, including hardware and software engineers who build, maintain, and operate automated laboratories and the software needed to learn from data and to design experiments.
- Fostering collaborative research that aims at developing and using ARWs and that facilitates sharing workflows, code, data, and data products in ways that respect and protect privacy considerations.

RECOMMENDATION 4: Culture and Incentives

Research funders, research institutions, and disciplines should work to create an automated research workflow (ARW)-friendly culture by making changes in incentive and reward structures aimed at encouraging behaviors that are central to realizing the potential of ARWs. These include

- Encouraging team science and multidisciplinary teams.
- Using funding support and provisions for data management plans to encourage development and curation of FAIR, responsible, and good-quality data resources.
- Developing, improving, and sharing software resources.
- Reporting reproducible results.
- Helping others adopt ARW practices.
- Pursuing international collaboration when possible in order to accelerate progress toward implementing the above changes at scale.

Finding D: Legal and Policy Issues

In addition to barriers to progress that exist within the research process itself, there are legal and policy issues that affect implementation of automated research workflows in specific domains that will require international multistakeholder efforts to address.

RECOMMENDATION 5: Preserving Privacy

Research enterprise funders, performers, publishers, and beneficiaries should work with governments, data privacy experts, and other entities to address the legal, policy, and associated technical barriers to implementing automated research workflows in use-inspired applications in specific domains and explore solutions to make the outputs available through privacy-preserving algorithms, federated learning approaches to using data, and other methods.

Bibliography

AGU (American Geophysical Union). 2021. Guidance for AGU authors—Jupyter Notebooks. Available at https://data.agu.org/resources/jupyter-notebooks-guidance. Accessed November 23, 2021.
AI/ML workflows on OpenShift. Red Hat OpenShift. Available at https://demo.openshift.com/en/latest/aiml-workflows/. Accessed June 2020.
Alberts, B. M. 2013. Impact factor distortions. *Science* 340(6134):787. doi: 10.1126/science.1240319.
Alberts, B. M., M. W. Kirschner, S. Tilghman, and H. Varmus. 2014. Rescuing US biomedical research from its systemic flaws. *Proceedings of the National Academy of Sciences of the United States of America* 111(16):5773–5777. doi: 10.1073/pnas.1404402111.
ALLEA. 2017. The European code of conduct for research integrity. Available at https://allea.org/code-of-conduct/. Accessed January 21, 2021.
Altintas, I. 2018. Evolving Role of Scientific Workflows in a Highly Networked, Collaborative and Dynamic Data-Driven World. Keynote Talk, Works 2018 Workshop, November 11, 2018. Available at http://works.cs.cf.ac.uk/2018/program.php. Accessed April 13, 2022.
Altintas, I. 2019. SC19: Next Generation Disaster Intelligence Using the Continuum of Computing and Data Technologies. Address at SC19, November 21, 2019. Available at https://wifire.ucsd.edu/node/115. Accessed April 13, 2022.
Altintas, I. 2020a. Challenges and Opportunities for Composable AI-Integrated Applications at the Digital Continuum: Keynote. 2020 IEEE/ACS 17th International Conference on Computer Systems and Applications (AICCSA). doi: 10.1109/AICCSA50499.2020.9316494.
Altintas, I. 2020b. Using dynamic data driven cyberinfrastructure for next generation disaster intelligence. In F. Darema, E. Blasch, S. Ravela, and A. Aved (Eds.). Dynamic Data Driven Applications Systems: Third International Conference, DDDAS 2020, Boston, MA, USA, October 2-4, 2020, Proceedings (18-21). Handel, Switzerland: Springer Nature Switzerland. doi: 0.1007/978-3-030-61725-7.
Altintas, I., J. Block, R. de Callafon, D. Crawl, C. Cowart, A. Gupta, M. Nguyen, H. W. Braun, J. Schulze, M. Gollner, A. Trouve, and L. Smarr. 2015. Towards an integrated cyberinfrastructure for scalable data-driven monitoring, dynamic prediction and resilience of wildfires. *Procedia Computer Science* 51:1633–1642. doi: 10.1016/j.procs.2015.05.296.

Altintas, I., S. Purawat, D. Crawl, A. Singh, and K. Marcus. 2019. Toward a methodology and framework for workflow-driven team science. *Computing in Science & Engineering* 21:37–48. doi: 10.48550/arXiv.1903.01403.

Anthony, K. 2020. New open release allows theorists to explore LHC data in a new way. CERN Accelerating Science. Available at https://home.cern/news/news/knowledge-sharing/new-open-release-allows-theorists-explore-lhc-data-new-way. Accessed January 9, 2020.

Aspuru-Guzik, A., and K. Persson. 2018. *Materials Acceleration Platform: Accelerating advanced energy materials discovery by integrating high-throughput methods and artificial intelligence. Mission Innovation: Innovation Challenge 6.* Available at http://nrs.harvard.edu/urn-3:HUL.InstRepos:35164974. Accessed June 17, 2020.

Atkinson, M., S. Gesing, J. Montagnat, and I. Taylor. 2017. Scientific workflows: Past, present, and future. *Future Generation Computer Systems* 75:216–227.

Aucamp, I. 2020. Computational data analysis workflow systems. Available at https://github.com/common-workflow-language/common-workflow-language/wiki/Existing-Workflow-systems. Accessed April 27, 2021.

Barber, G. 2019. Artificial intelligence faces reproducibility crisis. *Wired*. Available at https://www.wired.com/story/artificial-intelligence-confronts-reproducibility-crisis/. Accessed January 12, 2021.

Barga, R., and D. Gannon. 2007. Scientific versus business workflows. In *Workflows for e-science*, I. J. Taylor, E. Deelman, D. B.Gannon, and M. Shields (eds.). London: Springer. doi: 10.1007/978-1-84628-757-2_2.

Bauer, P., A. Thorpe, and G. Brunet. 2015. The quiet revolution of numerical weather prediction. *Nature* 525:47–55. doi: 10.1038/nature14956.

Beckman, P. 2020. Supporting tools and systems. Presentation at the Workshop on Opportunities for Accelerating Scientific Discovery: Realizing the Potential of Advanced and Automated Workflows, March 16–17.

Begley, C. G., and L. M. Ellis. 2012. Raise standards for preclinical cancer research. *Nature* 483:531–533. doi: 10.1038/483531a.

Bellissimo, J. 2019. Intelligent workflows 101: Revolutionizing the way your business works. *IBM Smarter Business Review*. Available at https://www.ibm.com/blogs/services/2019/04/29/intelligent-workflows-101-revolutionizing-the-way-your-business-works/. Accessed November 30, 2021.

Bergmann, U., L. Deer, and J. Langer. 2021. *Reproducible data analytic workflows for economics: An introduction to Snakemake.* Available at https://lachlandeer.github.io/snakemake-econ-r-tutorial/. Accessed April 14, 2021.

Bhattacharya, S. 2019. The new dawn of AI: Federated learning. *Towards Data Science*. January 27. Available at https://towardsdatascience.com/the-new-dawn-of-ai-federated-learning-8ccd9ed7fc3a. Accessed November 30, 2021.

Biemann, C., G. R. Crane, C. D. Fellbaum, and A. Mehler. 2014. Computational humanities—bridging the gap between computer science and digital humanities. *Dagstuhl Reports* 4(7):80–111. doi: 10.4230/DagRep.4.7.80.

Brennan, P., E. Green, and B. Tromberg. 2020. Concept clearance for artificial intelligence for biomedical excellence (AIBLE). Available at https://dpcpsi.nih.gov/sites/default/files/CoC_May_2020_1.05PM_CF_Concept_Clearance_AIBLE_Background_Brennan_508.pdf. Accessed April 14, 2021.

Bucci, E. 2018. Automatic detection of image manipulations in the biomedical literature. *Nature: Cell Death and Disease* 9:400. doi: 10.1038/s41419-018-0430-3.

Burgelman, J. 2020. Emerging policy (pre) conditions for research data management. Presentation at the Workshop on Opportunities for Accelerating Scientific Discovery: Realizing the Potential of Advanced and Automated Workflows, March 16–17.

Caliskan, A., J. Bryson, and A. Narayanan. 2017. Semantics derived automatically from language corpora contain human-like biases. *Science* 356:183–186. doi: 10.1126/science.aal4230.

Casadevall, A., and F. C. Fang. 2012. Reforming science: Methodological and cultural reforms. *Infection and Immunity* 80(3):891–896. doi: 10.1128/IAI.06183-11.

Cedeno-Mieles, V., Z. Hu, Y. Ren, X. Deng, N. Contractor, S. Ekanayake, J. Epstein, B. Goode, G. Korkmaz, C. Kuhlman, D. Machi, M. Macy, M. V. Marathe, N. Ramakrishnan, P. Saraf, and N. Self. 2020. Data analysis and modeling pipeline for controlled networked social science experiments. *PLoS ONE* 5(11):e0242453. doi: 10.1371/journal.pone.0242453.

Cernak, T. 2016. Synthesis in the chemical space age. *Chem* 1(1):6–9. doi: 10.1016/j.chempr.2016.06.002.

Cernak, T. 2020. Opportunities in automated synthesis of small molecules. Presentation at the Workshop on Opportunities for Accelerating Scientific Discovery: Realizing the Potential of Advanced and Automated Workflows, March 16–17.

Chen, X., S. Dallmeier-Tiessen, R. Dasler. S. Feger, P. Fokianos, J. B. Gonzalez, H. Hirvonsalo, D. Kousidis, A. Lavasa, S. Mele, D. R. Rodriguez, T. Šimko, T. Smith, A. Trisovic, A. Trzcinska, I. Tsanaktsidis, M. Zimmermann, K. Cranmer, L. Heinrich, G. Watts, M. Hildreth, L. L. Iglesias, K. Lassila-Perini, and S. Neubert. 2019. Open is not enough. *Nature Physics* 15:113–119. doi: 10.1038/s41567-018-0342-2.

Chetty, R., and J. Friedman. 2019. A practical method to reduce privacy loss when disclosing statistics based on small samples. *Journal of Privacy and Confidentiality* 9(2). doi: 10.29012/jpc.716.

Christian, T. M., S. Lafferty-Hess, W. G. Jacoby, and T. M. Carsey. 2018. Operationalizing the republication standard: A case study of the data curation and verification workflow for scholarly journals. Available at https://osf.io/preprints/socarxiv/cfdba/. Accessed March 18, 2021.

Cleary, E., A. Garbuno-Inigo, S. Lan, T. Schneider, and A. M. Stuart. 2021. Calibrate, emulate, sample. *Journal of Computational Physics* 424:109716. doi: 10.1016/j.jcp.2020.109716.

Cohen-Boulakia, S., K. Belhajjame, O. Collin, J. Chopard, C. Froidevaux, A. Gaignard, K. Hinsen, P. Larmande, Y. Le Bras, F. Lemoine, F. Mareuil, H. Ménager, C. Pradal, and C. Blanchet. 2017. Scientific workflows for computational reproducibility in the life sciences: Status, challenges and opportunities. *Future Generation Computer Systems* 75:284–298. doi: 10.1016/j.future.2017.01.012.

COPDESS (Coalition for Publishing Data in the Earth and Space Sciences). 2021. Enabling FAIR Data Project. Available at http://www.copdess.org/enabling-fair-data-project/. Accessed November 8, 2021.

Crane, G. 2020. Philosophy at scale. Presentation at the Workshop on Opportunities for Accelerating Scientific Discovery: Realizing the Potential of Advanced and Automated Workflows, March 16–17.

Cranmer, K. 2020. Accelerating physics with advanced, automated workflows. Presentation at the Workshop on Opportunities for Accelerating Scientific Discovery: Realizing the Potential of Advanced and Automated Workflows, March 16–17.

Cranmer, K., and I. Yavin. 2010. RECAST: Extending the impact of existing analyses. Preprint. Available at https://arxiv.org/abs/1010.2506. Accessed June 19, 2020.

Cranmer, K., J. Brehmer, and G. Louppec. 2020. The frontier of simulation-based inference. Preprint. Available at https://arxiv.org/pdf/1911.01429.pdf. Accessed June 19, 2020.

Crosas, M., G. King, J. Honaker, and L. Sweeney. 2015. Automating open science for big data. *Annals of the American Academy of Political and Social Science* 659(1):260–273. doi: 10.1177/0002716215570847.

CRS (Congressional Research Service). 2020. Federal research and development (R&D) funding: FY2021. December 17. Available at https://fas.org/sgp/crs/misc/R46341.pdf. Accessed May 15, 2021.

CSTCloud (China Science and Technology Cloud). 2020. About the CSTCloud. Available at https://www.cstcloud.net/cstcloud.htm. Accessed April 27, 2021.

CWL (*Common Workflow Language user guide*). 2020. Available at https://www.commonwl.org/user_guide/. Accessed June 2020.

Davidson, S. B., S. Khanna, S. Roy, and S. C. Boulakia. 2010. Privacy issues in scientific workflow provenance. Available at https://repository.upenn.edu/cis_papers/669. Accessed June 19, 2020.

de Carvalho, E. C. A., M. K. Jayanti, A. P. Batilana, A. M. O. Kozan, M. J. Rodrigues, J. Shah, M. R. Loures, S. Patil, P. Payne, and R. Pietrobon. 2010. Standardizing clinical trials workflow representation in UML for international site comparison. *PLoS ONE* 5(11):e13893. doi: 10.1371/journal.pone.0013893.

Deelman E., A. Mandal, and M. Jiang. 2019. The role of machine learning in scientific workflows. *International Journal of High Performance Computing Applications* 33(6):1128–1139. doi: 10.1177/1094342019852127.

Deelman E., R. Ferreira da Silva, K. Vahi, M. Rynge, R. Mayani, R. Tanaka, W. Whitcup, and M. Livny. 2020. The Pegasus workflow management system: Translational computer science in practice. *Journal of Computational Science* 52:101200. doi: 10.1016/j.jocs.2020.101200.

Dockstore. 2021. Dockstore: An app store for bioinformatics. Available at https://dockstore.org/. Accessed November 23, 2021.

DOE (U.S. Department of Energy). 2019. *Workshop report on basic research needs for scientific machine learning: Core technologies for artificial intelligence*. Office of Scientific and Technical Information. Available at https://www.osti.gov/servlets/purl/1478744. Accessed April 4, 2021.

DORA (Declaration on Research Assessment). 2013. San Francisco declaration on research assessment. Available at https://sfdora.org/read. Accessed January 12, 2021.

Duke University. 2021. Data management plan. Duke University Office of Scientific Integrity. Available at https://dosi.duke.edu/advancing-scientific-integrity-services-and-training/accountability-research/data-management-plan. Accessed November 15, 2021.

EC (European Commission). 2016. Realising the European Open Science Cloud: First report and recommendations. Available at https://ec.europa.eu/research/openscience/pdf/realising_the_european_open_science_cloud_2016.pdf. Accessed June 19, 2020.

EC. 2019. Cost-benefit analysis for FAIR research data. Available at https://op.europa.eu/en/publication-detail/-/publication/d375368c-1a0a-11e9-8d04-01aa75ed71a1. Accessed June 19, 2020.

EC. 2020. *Progress on open science: Towards a shared research knowledge system: Final report of the open science policy platform*. Available at https://data.europa.eu/doi/10.2777/00139. Accessed January 12, 2021.

EC. 2021a. Artificial intelligence (AI): Artificial intelligence research, funding, policy, and related publications. Available at https://ec.europa.eu/info/research-and-innovation/research-area/industrial-research-and-innovation/key-enabling-technologies/artificial-intelligence-ai_en. Accessed January 21, 2021.

EC. 2021b. European Open Science Cloud. Available at https://ec.europa.eu/info/research-and-innovation/strategy/goals-research-and-innovation-policy/open-science/european-open-science-cloud-eosc_en. Accessed January 12, 2021.

Einav, L., and J. Levin. 2014. Economics in the age of big data. *Science* 346(6210). doi: 10.1126/science.1243089.

EOSC (European Open Science Cloud). 2020. SRIA (Strategic Research and Innovation Agenda) of the European Open Science Cloud (EOSC). Available at https://www.eoscsecretariat.eu/sites/default/files/eosc-sria-v09.pdf. Accessed July 19, 2021.

EU (European Union). 2018. General data protection regulation (GDPR). Available at https://gdpr-info.eu/. Accessed December 10, 2021.

European Parliament Panel for the Future of Science and Technology. 2019. How the general data protection regulation changes the rules of scientific research. Available at https://www.europarl.europa.eu/RegData/etudes/STUD/2019/634447/EPRS_STU(2019)634447_EN.pdf. Accessed June 19, 2020.

Feingenbuam, J., and D. J. Weitzner. 2018. On the incommensurability of laws and technical mechanisms: Or, what cryptography can't do. In *Security Protocols 2018. Lecture Notes in Computer Science*, V. Matyáš, P. Švenda, F. Stajano, B. Christianson, and J. Anderson (eds.). Cham: Springer, pp. 266-279. Available at https://link.springer.com/chapter/10.1007/978-3-030-03251-7_31#citeas. Accessed June 19, 2020.

Ferreira da Silva, R., H. Casanova, K. Chard, T. Coleman, D. Laney, D. Ahn, S. Jha, D. Howell, S. Soiland-Reyes, I. Altintas, D. Thain, R. Filgueira, Y. Babuji, R. Badia, B. Balis, S. Caino-Lores, S. Callaghan, F. Coppens, M. Crusoe, K. De, F. Di Natale, T. M. A. Do; B. Enders, T. Fahringer, A. Fouilloux, G. Fursin, A. Gaignard, A. Ganose, D. Garijo, S. Gesing, C. Goble, A. Hassan, S. Huber, D. S. Katz, U. Leser, D. Lowe, B. Ludascher, K. Maheshwari, M. Malawski, R. Mayani, K. Mehta, A. Merzky, T. Munson, J. Ozik, L. Pottier, S. Ristov, M. Roozmeh, R. Souza, F. Suter, B. Tovar, M. Turilli, K, Vahi, A. Vidal-Torreira, W. Witcup, M. Wilde, A. Williams, M. Wolf, and J. Wozniak. 2021a. *Workflows Community Summit: Advancing the state-of-the-art of scientific workflows management systems research and development.* Technical Report. Zenodo. doi: 10.5281/zenodo.4915801.

Ferreira da Silva, R., H. Casanova, K. Chard, D. Laney, D. Ahn, S. Jha, C. Goble, L. Ramakrishnan, L. Peterson, B. Enders, D. Thain, I. Altintas, Y. Babuji, R. Badia, V. Bonazzi, T. Coleman, M. Crusoe, E. Deelman, F. Di Natale, P. Di Tommaso, T. Fahringer, R. Filgueira, G. Fursin, A. Ganose, B. Gruning, D. S. Katz, O. Kuchar, A. Kupresanin, B. Ludascher, K. Maheshwari, M. Mattoso, K. Mehta, T. Munson, J. Ozik, T. Peterka, L. Pottier, T. Randles, S. Soiland-Reyes, B. Tovar, M. Turilli, T. Uram, K. Vahi, M. Wilde, M. Wolf, and J. Wozniak. 2021b. *Workflows Community Summit: Bringing the scientific workflows community together.* Technical Report. Zenodo. doi: 10.5281/zenodo.4606958.

Fox, G. 2020. Status and trajectory of supporting tools and systems. Presentation at the Workshop on Opportunities for Accelerating Scientific Discovery: Realizing the Potential of Advanced and Automated Workflows, March 16–17.

Freedman, A. 2019. Weather is turning into big business. And that could be trouble for the public. *Washington Post.* Available at https://www.washingtonpost.com/business/2019/11/25/weather-is-big-business-its-veering-toward-collision-with-federal-government. Accessed April 14, 2021.

Galaxy. 2021a. 30,000 users. Blog post. Available at https://galaxyproject.eu/posts/2021/04/25/30000user/. Accessed December 10, 2021.

Galaxy. 2021b. Global platform for the analysis of SARS-CoV-2 data: Genomics, cheminformatics, and proteomics. 2021. Available at https://covid19.galaxyproject.org/. Accessed November 23, 2021.

Gardner, T. 2020. Presentation at the Workshop on Opportunities for Accelerating Scientific Discovery: Realizing the Potential of Advanced and Automated Workflows, March 16–17.

Gil, Y., D. Garijo, V. Ratnakar, R. Mayani, R. Adusumilli, H. Boyce, A. Srivastava, and P. Mallick. 2017. Towards continuous scientific data analysis and hypothesis evolution. *Proceedings of the AAAI Conference on Artificial Intelligence* 31(1). Available at https://ojs.aaai.org/index.php/AAAI/article/view/11157. Accessed May 21, 2021.

Gil, Y., S. A. Pierce, H. Babaie, A. Banerjee, K. Borne, G. Bust, M. Cheatham, I. Ebert-Uphoff, C. Gomes, M. Hill, J. Horel, L. Hsu, J. Kinter, C. Knoblock, D. Krum, V. Kumar, P. Lermusiaux, Y. Liu, C. North, V. Pankratius, S. Peters, B. Plale, A. Pope, S. Ravela, J. Restrepo, A. Ridley, H. Samet, S. Shekhar, K. Skinner, P. Smyth, B. Tikoff, L. Yarmey, and J. Zhang. 2019. Intelligent systems for geosciences. *Communications of the ACM* 62(1). doi: 10.1145/3192335.

Gillespie, T., P. J. Boczkowski, and K. A. Foot. 2014. *Media technologies: Essays on communication, materiality, and society.* Cambridge, MA: MIT Press.

GitHub. 2021. Existing workflow systems. Available at https://github.com/common-workflow-language/common-workflow-language/wiki/Existing-Workflow-systems. Accessed September 9, 2020.

Global Commission on Adaptation. 2019. *Adapt now: A global call for leadership on climate resilience.* Available at https://gca.org/wp-content/uploads/2019/09/GlobalCommission_Report_FINAL.pdf. Accessed August 23, 2021.

GO FAIR. 2016. FAIR Principles. Available at https://www.go-fair.org/fair-principles. Accessed June 19, 2020.

Goble, C., S. Cohen-Boulakia, S. Soiland-Reyes, D. Garijo, Y. Gil, M. R. Crusoe, K. Peters, and D. Schober. 2020. FAIR computational workflows. *Data Intelligence* 2(1-2):108–121. doi: 10.1162/dint_a_00033.

Google. 2021. Artificial intelligence at Google: Our principles. Available at https://ai.google/principles/. Accessed November 29, 2021.

Google Cloud. 2020. Introduction to AI Platform. Available at https://cloud.google.com/ai-platform/docs/technical-overview. Accessed June 22, 2020.

Granger, B. 2020. Status and Trajectory of Supporting Tools and Systems. Presentation at the Workshop on Opportunities for Accelerating Scientific Discovery: Realizing the Potential of Advanced and Automated Workflows, March 16–17.

Hacking, I. 1983. *Representing and intervening: Introductory topics in the philosophy of natural science.* Cambridge University Press.

Hardisty, A., and P. Wittenburg. 2020. Canonical Workflow Framework for Research (CWFR), version 2: Position Paper. Working Paper. December. Center for Open Science. https://osf.io/9e3vc/.

Hattrick-Simpers, J. 2020. How robots could teach us to trust AI. NIST Taking Measure blog. Available at https://www.nist.gov/blogs/taking-measure/how-robots-could-teach-us-trust-ai. Accessed June 18, 2020.

Hein Lab. 2020. Designing and applying in situ analysis to enable chemical discovery. Available at https://groups.chem.ubc.ca/jheints1. Accessed June 18, 2020.

Hepworth, K. J., and C. Church. 2018. Racism in the machine: Visualization ethics in the digital humanities. *Digital Humanities Quarterly* 12(4). Available at http://www.digitalhumanities.org/dhq/vol/12/4/000408/000408.html. Accessed June 22, 2020.

Hey, T., K. Butler, S. Jackson, and J. Thiyagalingam. 2020. The Royal Society Publishing. Available at https://doi.org/10.1098/rsta.2019.0054. Accessed November 21, 2020.

Hill, A. C. 2021. COVID's lesson for climate research: Go local. *Nature* 595:9. doi: 10.1038/d41586-021-01747-9.

Hill, J., G. Mulholland, K. Persson, R. Seshadri, C. Wolverton, and B. Meredig. 2016. Materials science with large-scale data and informatics: Unlocking new opportunities. *MRS Bulletin* 41:399–409. Available at https://perssongroup.lbl.gov/papers/hill2016-mrsbull.pdf. Accessed June 18, 2020.

Hinsen, K. 2019. Dealing with software collapse. *Computing in Science & Engineering* 21(3):104-108. doi: 10.1109/MCSE.2019.2900945.

Holdren, J. P. 2013. Increasing access to the results of federally funded scientific research. Memorandum to Heads of Executive Departments and Agencies. Washington, DC: Office of Science and Technology Policy.

Hunt, E. 2016. Tay, Microsoft's AI chatbot, gets a crash course in racism from Twitter. *The Guardian*. Available at https://www.theguardian.com/technology/2016/mar/24/tay-microsofts-ai-chatbot-gets-a-crash-course-in-racism-from-twitter. Accessed June 22, 2020.

Hyysalo, J., M. Oivo, and P. Kuvaja. 2017. A design theory for cognitive workflow systems. *International Journal of Software Engineering and Knowledge Engineering* 27(1):125–151. doi: 10.1142/S0218194017500061.

IBM. 2021. Workflow. Available at https://www.ibm.com/cloud/learn/workflow. Accessed December 3, 2021.

IEEE. 2020. IEEE 2791-2020—IEEE standard for bioinformatics analyses generated by high-throughput sequencing (HTS) to facilitate communication. Available at https://standards.ieee.org/standard/2791-2020.html. Accessed November 24, 2021.

IMI (Innovative Medicines Initiative). 2021. MELLODDY project factsheet. Available at https://www.imi.europa.eu/projects-results/project-factsheets/melloddy. Accessed November 30, 2021.

Institute for Ethical AI & Machine Learning. 2021. The responsible machine learning principles. Available at https://ethical.institute/principles.html. Accessed November 29, 2021.

Ioannidis, J. P. A., K. W. Boyack, H. Small, A. A. Sorensen, and R. Klavans. 2014. Bibliometrics: Is your most cited work your best? *Nature* 514:561–562. doi: 10.1038/514561a.

IPCC (Intergovernmental Panel on Climate Change). 2021. *Climate change 2021: The physical science basis. Contribution of Working Group I to the Sixth Assessment Report of the Intergovernmental Panel on Climate Change*, V. Masson-Delmotte, P. Zhai, A. Pirani, S. L. Connors, C. Péan, S. Berger, N. Caud, Y. Chen, L. Goldfarb, M. I. Gomis, M. Huang, K. Leitzell, E. Lonnoy, J. B. R. Matthews, T. K. Maycock, T. Waterfield, O. Yelekçi, R. Yu, and B. Zhou (eds.). Cambridge University Press.

Jones, S. 2021. A typology of the components of the Global Open Research Commons. Research Data Alliance. Available at https://www.rd-alliance.org/plenaries/rda-17th-plenary-meeting-edinburgh-virtual/typology-components-global-open-research. Accessed November 20, 2020.

Juric, M., E. Bellm, and L. Guy. 2019. Machine learning applications with LSST: From data processing to knowledge discovery. *American Astronomical Society Meeting Abstracts No. 233*. Available at https://ui.adsabs.harvard.edu/abs/2019AAS...23312601J/abstract. Accessed June 22, 2020.

Kahkoska, A. R., T. J. Abrahamsen, G. C. Alexander, T. D. Bennett, C. G. Chute, M. A. Haendel, K. R. Klein, H. Mehta, J. D. Miller, R. A. Moffitt, T. Stürmer, K. Kvist, J. B. Buse, and N3C Consortium. 2021. Association between glucagon-like peptide 1 receptor agonist and sodium–glucose cotransporter 2 inhibitor use and COVID-19 outcomes. *Diabetes Care* 44(7):1564–1572. doi: 10.2337/dc21-0065.

Kalnay, E. 2003. *Atmospheric modeling, data assimilation and predictability*. Cambridge, UK: Cambridge University Press.

Kangas, J. D., A. W. Naik, and R. F. Murphy. 2014. Efficient discovery of responses of proteins to compounds using active learning. *BMC Bioinformatics* 15(143). doi: 10.1186/1471-2105-15-143.

Kim, B., M. Wattenberg, J. Gilmer, C. J. Cai, J. Wexler, F. Viegas, and R. A. Sayres. 2018. Interpretability beyond feature attribution: Quantitative testing with concept activation vectors (TCAV). In *Proceedings of the 35th International Conference on Machine Learning, Stockholm, Sweden*. PMLR 80:2668–2677. Available at https://research.google/pubs/pub47077. Accessed June 22, 2020.

Kizilcec, R., J. Reich, M. Yeomans, C. Dann, E. Brunskill, G. Lopez, S. Turkay, J. J. Williams, and D. Tingley. 2020. Scaling up behavioral science interventions in online education. *Proceedings of the National Academy of Sciences of the United States of America* 117(26):14900–14905. doi: 10.1073/pnas.1921417117.

Koolen, M., S. Kumpulainen, and L. Melgar-Estrada. 2020. A workflow analysis perspective to scholarly research tasks. In *2020 Conference on Human Information Interaction and Retrieval (CHIIR '20)*. March 14–18, 2020, Vancouver, BC, Canada. doi: 10.1145/3343413.3377969.

Kusnezov, D. 2020. National Academies workflow discussion. Presentation at the Workshop on Opportunities for Accelerating Scientific Discovery: Realizing the Potential of Advanced and Automated Workflows, March 16–17.

Lane, J. 2020. Status and trajectory of supporting tools and systems. Presentation at the Workshop on Opportunities for Accelerating Scientific Discovery: Realizing the Potential of Advanced and Automated Workflows, March 16–17.

Lane, J., I. Mulvany, and P. Nathan. 2020. *Rich search and discovery for research datasets: Building the next generation of scholarly infrastructure*. Sage Press. Available at https://wagner.nyu.edu/impact/research/publications/rich-search-and-discovery-for-research-datasets-building-next. Accessed April 14, 2021.

Lawrence, P. A. 2007. The mismeasurement of science. *Current Biology* 17:R583–R585. doi: 10.1016/j.cub.2007.06.014.

L'Heureux, A., K. Grolinger, H. F. Elyamany, and M. A. M. Capretz. 2017. Machine learning with big data: Challenges and approaches. *IEEE Access* 5:7776–7797. doi: 10.1109/ACCESS.2017.2696365.

Maier, W., S. Bray, M. van den Beek, D. Bouvier, N. Coraor, M. Miladi, B. Singh, J. Rambla De Argila, D. Baker, N. Roach, S. Gladman, F. Coppens, D. P. Martin, A. Lonie, B. Grüning, S. L. Kosakovsky Pond, and A. Nekrutenko. 2021. Freely accessible ready to use global infrastructure for SARS-CoV-2 monitoring. *bioRxiv* (preprint). doi: 10.1101/2021.03.25.437046.

Martinez, P. A. 2021. FAIR principles for research software (FAIR4RS principles). Available at https://rd-alliance.org/group/fair-research-software-fair4rs-wg/outcomes/fair-principles-research-software-fair4rs. Accessed July 19, 2021.

McNutt, M. 2017. Convergence in the geosciences. *GeoHealth* 1:2–3. doi: 10.1002/2017GH000068.

McNutt, M., M. Bradford, J. M. Drazen, B. Hanson, B. Howard, K. H. Jamieson, V. Kiermer, E. Marcus, B. K. Pope, R. Schekman, S. Swaminathan, P. J. Stang, and I. M. Verma. 2018. *Proceedings of the National Academy of Sciences of the United States of America* 115(11):2557–2560. doi: 10.1073/pnas.1715374115.

McPhillips, T., T. Song, T. Kolisnik, S. Aulenbach, K. Belhajjame, K. Bocinsky, Y. Cao, F. Chirigati, S. Dey, J. Freire, D. Huntzinger, C. Jones, D. Koop, P. Missier, M. Schildhauer, C. Schwalm, Y. Wei, J. Cheney, M. Bieda, and B. Ludaescher. 2015. YesWorkflow: A user-oriented, language-independent tool for recovering workflow information from scripts. *arXiv*:1502.02403. doi: 10.48550/arXiv.1502.02403.

McQueen, T. 2020. Materials discovery. Presentation at the Workshop on Opportunities for Accelerating Scientific Discovery: Realizing the Potential of Advanced and Automated Workflows, March 16–17.

Mehra, M. R., S. S. Desai, F. Ruschitzka, and A. N. Patel. 2020. RETRACTED: Hydroxychloroquine or chloroquine with or without a macrolide for treatment of COVID-19: A multinational registry analysis. *The Lancet*. doi: 10.1016/S0140-6736(20)31180-6.

Mello, M. M., G. Triantis, R. Stanton, E. Blumenkranz, and D. M. Studdert. 2020. Waiting for data: Barriers to executing data use agreements. *Science* 367(6474):150–152. doi: 10.1126/science.aaz7028.

Mons, B. 2020. Invest 5% of research funds in ensuring data are reusable. *Nature* 578(491). doi: https://doi.org/10.1038/d41586-020-00505-7

Moundas, C., and D. Peloquin. 2020. Insight: California bill clarifies privacy law's ambiguities for medical, research communities. *Bloomberg Law*. Available at https://news.bloomberglaw.com/health-law-and-business/insight-california-bill-clarifies-privacy-laws-ambiguities-for-medical-research-communities. Accessed June 22, 2020.

Murphy, R. F. 2011. An active role for machine learning in drug development. *Nature Chemical Biology* 7(6):327–330. doi: 10.1038/nchembio.576.

Murphy, R. F. 2020. Self-driving instruments: The need for closed loop, AI-driven biomedical research. Presentation at the Workshop on Opportunities for Accelerating Scientific Discovery: Realizing the Potential of Advanced and Automated Workflows, March 16–17.

Naik, A. W., J. D. Kangas, D. P. Sullivan, and R. F. Murphy. 2016. Active machine learning-driven experimentation to determine compound effects on protein patterns. *eLife* 5. doi: 10.7554/eLife.10047.

Nangia, U., and D. S. Katz. 2017. Understanding software in research: Initial results from examining nature and a call for collaboration. In *IEEE 13th International Conference on e-Science (e-Science)*, pp. 486–487. doi: 10.1109/eScience.2017.78.

NAS (National Academy of Sciences). 2018. *The frontiers of machine learning: 2017 Raymond and Beverly Sackler U.S.-U.K. scientific forum*. Washington, DC: The National Academies Press. doi: 10.17226/25021.

NAS, NAE, and IOM (National Academy of Sciences, National Academy of Engineering, and Institute of Medicine). 2009. *Ensuring the integrity, accessibility, and stewardship of research data in the digital age*. Washington, DC: The National Academies Press. doi: 10.17226/12615.

NASEM (National Academies of Sciences, Engineering, and Medicine). 2012. *A national strategy for advancing climate modeling*. Washington, DC: The National Academies Press. Doi: 10.17226/13430.

NASEM. 2015. *Enhancing the effectiveness of team science*. Washington, DC: The National Academies Press. doi: 10.17226/19007.

NASEM. 2017. *Fostering integrity in research*. Washington, DC: The National Academies Press. doi: 10.17226/21896.

NASEM. 2018a. *Artificial intelligence and machine learning to accelerate translational research*. Washington, DC: The National Academies Press.

NASEM. 2018b. *Open science by design: Realizing a vision for 21st century research*. Washington, DC: The National Academies Press. doi: 10.17226/25116.

NASEM. 2018c. *The frontiers of machine learning: 2017 Raymond and Beverly Sackler U.S.-U.K. scientific forum*. Washington, DC: The National Academies Press. doi: 10.17226/25021.

NASEM. 2019a. *Frontiers of materials research: A decadal survey*. Washington, DC: The National Academies Press. doi: 10.17226/25244.

NASEM. 2019b. *Reproducibility and replicability in science.* Washington, DC: The National Academies Press. doi: 10.17226/25303.

NASEM. 2020a. *Long-term use of biomedical research.* Washington, DC: The National Academies Press. doi: 10.17226/25653.

NASEM. 2020b. *Neuroscience data in the cloud.* Washington, DC: The National Academies Press. doi: 10.17226/25653.

NASEM. 2021. *Developing a toolkit for fostering open science practices: Proceedings of a workshop.* Washington, DC: The National Academies Press. doi: 10.17226/26308.

Nathan, P. 2020. The future of AI in rich context. Chapter 12 in *Rich search and discovery for research datasets: Building the next generation of scholarly infrastructure,* J. Lane, I. Mulvany, and P. Nathan (eds). Sage Press.

Nature. 2017. Integrity starts with the health of research groups. 545:5–6. Available at https://www.nature.com/news/integrity-starts-with-the-health-of-research-groups-1.21921. Accessed January 13, 2021.

Nature. 2021. Reporting standards and availability of data, materials, code and protocols. Available at https://www.nature.com/nature-research/editorial-policies/reporting-standards. Accessed February 13, 2021.

Nature Physics. 2019. A problem shared is a problem halved (editorial). 15(107). doi: 10.1038/s41567-019-0434-7.

NCATS (National Center for Advancing Translational Sciences). 2021. About the National COVID Cohort Collaborative. National Institutes of Health. Available at https://ncats.nih.gov/n3c/about. Accessed January 13, 2021.

NIH (National Institutes of Health). 2018. NIH Data Commons Pilot Phase Consortium. Available at https://commonfund.nih.gov/commons/awardees. Accessed August 21, 2021.

NIH. 2020. Artificial Intelligence for Biomeical Excellence (AIBLE). Available at https://dpcpsi.nih.gov/sites/default/files/CoC_May_2020_1.05PM_Concept_Clearance_AIBLE_Brennan_508.pdf. Accessed April 4, 2020.

NIH. 2021. Common Fund Homepage. Available at https://commonfund.nih.gov. Accessed April 14, 2021.

NISO (National Information Standards Organization). 2021. Reproducibility badging and definitions. Available at https://www.niso.org/publications/rp-31-2021-badging. Accessed November 20, 2020.

NIST (National Institute of Standards and Technology). 2021. Research Data Framework (RDaF). Available at https://www.nist.gov/programs-projects/research-data-framework-rdaf. Accessed June 3, 2020.

NITRD (Networking and Information Technology Research and Development Program). 2020. Pioneering the future advanced computing ecosystem: A strategic plan. Available at https://www.nitrd.gov/pubs/Future-Advanced-Computing-Ecosystem-Strategic-Plan-Nov-2020.pdf. Accessed January 18, 2021.

NSF (National Science Foundation). 2007. *Cyberstructure vision for 21st century discovery.* Arlington, VA: National Science Foundation. Available at https://www.nsf.gov/pubs/2007/nsf0728/nsf0728.pdf. Accessed June 22, 2020.

NSF. 2020a. NSF big idea: Growing convergence research. Available at https://www.nsf.gov/od/oia/convergence/additional-resources/GCR-Powerpoint-Webinar-Jan-2020.pdf. Accessed November 15, 2020.

NSF. 2020b. NSF's ten big ideas. Available at https://www.nsf.gov/news/special_reports/big_ideas. Accessed June 22, 2020.

NSF. 2020c. Workshop on Smart Cyberinfrastructure. Available at http://smartci.sci.utah.edu/. Accessed February 20, 2021.

NSTL (National Science & Technology Council). 2019. *National strategic computing initiative update: Pioneering the future of computing.* Available at https://www.whitehouse.gov/wp-content/uploads/2019/11/National-Strategic-Computing-Initiative-Update-2019.pdf. Accessed June 22, 2020.

Nugent, R. 2020. Automating data science: Think about the human-machine interface. Presentation at the Workshop on Opportunities for Accelerating Scientific Discovery: Realizing the Potential of Advanced and Automated Workflows, March 16–17.

O'Brien, D. T., D. Offenhuber, J. Baldwin-Philippi, M. Sands, and E. Gordon. 2017. Uncharted territoriality in coproduction: The motivations for 311 reporting. *Journal of Public Administration Research & Theory* 27:320–335. doi: 10.1093/jopart/muw046.

ODSC Community (Open Data Science Conference Community). 2021. Building a robust data pipeline with the "dAG Stack": dbt, Airflow, and Great Expectations. Blog post. March 1. Available at https://opendatascience.com/building-a-robust-data-pipeline-with-the-dag-stack-dbt-airflow-and-great-expectations/. Accessed April 19, 2022.

OECD (Organisation for Economic Co-operation and Development). 2020. Building digital workforce capacity and skills for data-intensive science. *OECD Science, Technology, and Innovation Policy Papers* 90. Available at https://www.oecd-ilibrary.org/science-and-technology/building-digital-workforce-capacity-and-skills-for-data-intensive-science_e08aa3bb-en. Accessed July 19, 2021.

O'Hara, A. 2020. Model data use agreements: A practical guide. In *Handbook on using administrative data for research and evidence-based policy,* S. Cole, I. Dhaliwal, A. Sautmann, and L. Vilhuber (eds). Abdul Latif Jameel Poverty Action Lab. Available at https://admindatahandbook.mit.edu/book/v1.0-rc3/dua.html. Accessed December 12, 2020.

OSTP (Office of Science and Technology Policy). 2020. Public responses received for request for information 85 FR 3085: Draft desirable characteristics of repositories for managing and sharing data resulting from federally funded research. Available at https://www.whitehouse.gov/wp-content/uploads/2017/11/Desirable-Characteristics-RFC-Comments.pdf. Accessed November 17, 2021.

OSTP. 2021. The Biden Administration launches the National Artificial Intelligence Research Resource Task Force. Press release. Available at https://www.whitehouse.gov/ostp/news-updates/2021/06/10/the-biden-administration-launches-the-national-artificial-intelligence-research-resource-task-force/. Accessed March 3, 2022.

Owens, B. 2011. Reliability of 'new drug target' claims called into question. *Nature News Blog.* Available at http://blogs.nature.com/news/2011/09/reliability_of_new_drug_target.html. Accessed November 14, 2020.

Perkel, J. M. 2019. Workflow systems turn raw data into scientific knowledge. *Nature* 573:149–150. doi: 10.1038/d41586-019-02619-z.

Persson, K. 2020a. Conversation with Shreyas Cholia and Tapio Schneider, May 27, 2020.

Persson, K. 2020b. Making a material world better, faster now: Q&A with materials project director Kristin Persson. Available at https://eta.lbl.gov/news/article/making-material-world-better-faster. Accessed June 17, 2020.

Pew Research Center. 2019. Climate change still seen as the top global threat, but cyberattacks a rising concern. Findings of Pew Research Center survey. Available at https://www.pewresearch.org/global/2019/02/10/climate-change-still-seen-as-the-top-global-threat-but-cyberattacks-a-rising-concern. Accessed June 22, 2020.

Pfeiffer, J. K., and T. S. Dermody. 2021. Are too many scientists studying COVID? *Knowable Magazine.* Available at https://knowablemagazine.org/article/health-disease/2021/are-too-many-scientists-studying-covid. Accessed November 14, 2020.

Plale, B. 2020. AI and workflows for accelerated science. Presentation at the Workshop on Opportunities for Accelerating Scientific Discovery: Realizing the Potential of Advanced and Automated Workflows, March 16–17.

Plemmons, D. K., E. N. Baranski, K. Harp, D. D. Lo, C. K. Soderberg, T. M. Errington, B. A. Nosek, and K. M. Esterling. 2020. A randomized trial of a lab-embedded discourse intervention to improve research ethics. *Proceedings of the National Academy of Sciences of the United States of America* 117(3):1389–1394. doi: 10.1073/pnas.1917848117.

Project Jupyter. 2020. Estimate of public Jupyter Notebooks on GitHub. Available at https://nbviewer.jupyter.org/github/parente/nbestimate/blob/master/estimate.ipynb. Accessed June 3, 2020.

Queralt-Rosinach, N., R. Kaliyaperumal, C. Bernabé, Q. Long, S. A. Joosten, H. Jan van der Wijk, E. L. A. Flikkenschild, K. Burger, A. Jacobsen, B. Mons, M. Roos, BEAT-COVID Group, COVID-19 LUMC Group. 2021. Applying the FAIR principles to data in a hospital: Challenges and opportunities in a pandemic. *medRxiv* 2021.08.13.21262023 (Preprint). doi: 0.1101/2021.08.13.21262023.

Rasp, S., M. S. Pritchard, and P. Gentine. 2018. Deep learning to represent subgrid processes in climate models. *Proceedings of the National Academy of Sciences of the United States of America* 115(39):9684–9689. doi: 10.1073/pnas.1810286115.

Red Hat OpenShift. 2020. AI/ML Workflows on OpenShift. Demonstration Video. Available at https://demo.openshift.com/en/latest/aiml-workflows/. Accessed April 19, 2022.

RDA (Research Data Alliance). 2020. RDA COVID-19 guidelines and recommendations (draft versions). Available at https://doi.org/10.15497/RDA00046. Accessed September 7, 2021.

RDA. 2021a. Defining FAIR for machine learning (ML). Available at https://www.rd-alliance.org/defining-fair-machine-learning-ml. Accessed August 23, 2021.

RDA. 2021b. FAIR Principles for Research Software (FAIR4RS Principles). Available at https://doi.org/10.15497/RDA00065. Accessed August 23, 2021.

RDA-CODATA Legal Interoperability Interest Group. 2016. Legal interoperability of research data: Principles and implementation guidelines. Zenodo. http://doi.org/10.5281/zenodo.162241.

Reinsel, D., J. Gantz, and J. Rydning. 2018. The digitization of the world: From edge to core. IDC White Paper. Available at https://www.seagate.com/files/www-content/our-story/trends/files/idc-seagate-dataage-whitepaper.pdf. Accessed June 22, 2020.

Reiter, T., P. T. Brooks, L. Irber, S. E. K. Joslin, C. M. Reid, C. Scott, C. T. Brown, and N. T. Pierce-Ward. 2021. Streamlining data-intensive biology with workflow systems. *Gigascience* 10(1):giaa140. doi: 10.1093/gigascience/giaa140.

Retraction Watch. 2021. Retracted coronavirus (COVID-19) papers. Available at https://retractionwatch.com/retracted-coronavirus-covid-19-papers/. Accessed June 3, 2020.

Rothstein, M. A., and S. A. Tovino. 2019. Privacy risks of interoperable electronic health records: Segmentation of sensitive information will help. *Journal of Law, Medicine & Ethics* 47(4):771–777. doi:10.1177/1073110519897791.

Royal Society and Alan Turing Institute. 2019. The AI revolution in scientific research. Discussion paper. Available at https://royalsociety.org/-/media/policy/projects/ai-and-society/AI-revolution-in-science.pdf. Accessed December 7, 2021.

Rubin, R. 2020. NIH launches platform to serve as depository for COVID-19 medical data. *Journal of the American Medical Association* 324(4):326. doi: 10.1001/jama.2020.12646.

Rudin, C., and J. Radin. 2019. Why are we using black box models in AI when we don't need to? A lesson from an explainable AI competition. *Harvard Data Science Review* 1(2). doi: 10.1162/99608f92.5a8a3a3d.

Russell, A. 2020. Remarks at the Workshop on Opportunities for Accelerating Scientific Discovery: Realizing the Potential of Advanced and Automated Workflows, March 16–17.

Sansone, S. A., P. McQuilton, P. Rocca-Serra, A. Gonzalez-Beltran, M. Izzo, A. L. Lister, and M. Thurston. 2019. FAIRsharing as a community approach to standards, repositories and policies. *Nature Biotechnology* 37:358–367. doi: 10.1038/s41587-019-0080-8.

Schneider, T., S. Lan, A. Stuart, and J. Teixeira. 2017a. Earth system modeling 2.0: A blueprint for models that learn from observations and targeted high-resolution simulations. *Geophysical Research Letters* 44:12396–12417. doi: 10.1002/2017GL076101.

Schneider, T., J. Teixeira, C. S. Bretherton, F. Brient, K. G. Pressel, C. Schär, and A. P. Siebesma. 2017b. Climate goals and computing the future of clouds. *Nature Climate Change* 7:3–5. doi: 10.1038/nclimate3190.

Scholtens, S., M. Jetten, J. Böhmer, C. Staiger, I. Slouwerhof, M. van der Geest, and C.W.G. van Gelder. 2019. Final report: Towards FAIR data steward as profession for the lifesciences. Report of a ZonMw funded collaborative approach built on existing expertise. Zenodo. doi:10.5281/zenodo.3471707.

Service, R. F. 2019. AIs direct search for materials breakthroughs. *Science* 366(6471):1295–1296. doi: 10.1126/science.366.6471.1295.

Sharafeldin, N., B. Bates, Q. Song, V. Madhira, Y. Yan, S. Dong, E. Lee, N. Kuhrt, Y. R. Shao, F. Liu, T. Bergquist, J. Guinney, J. Su, and U. Topaloglu. 2021. Outcomes of COVID-19 in patients with cancer: Report from the National COVID Cohort Collaborative (N3C). *Journal of Clinical Oncology* 39(20):2232–2246. doi: 10.1200/JCO.21.01074.

Siebesma, A. P., C. S. Bretherton, A. Brown, A. Chlond, J. Cuxart, P. G. Duynkerke, and D. E. Stevens. 2003. A large eddy simulation intercomparison study of shallow cumulus convection. *Journal of Atmospheric Science* 60(10):1201–1219. doi: 10.1175/1520-0469(2003)60<1201:ALESIS>2.0.CO;2.

Smithies, J., C. Westling, A-M. Sichani, P. Mellen and A. Ciula. 2019. Managing 100 Digital Humanities Projects: Digital scholarship and archiving in King's Digital Lab. *Digital Humanities Quarterly* 13(1). Available at http://www.digitalhumanities.org/dhq/vol/13/1/000411/000411.html. Accessed April 19, 2022.

Somnath, S., C. R. Smith, N. Laanait, R. K. Vasudevan, A. Levlev, A. Belianinov, A. R. Lubini, M. Shankar, S. V. Kalinin, and S. Jesse. 2019. USID and pycroscopy—Open frameworks for storing and analyzing spectroscopic and imaging data. *arXiv* doi: 10.48550/ARXIV.1903.09515.

Sonntag, M., D. Karastoyanova, and E. Deelman. 2010. Bridging the gap between business and scientific workflows: Humans in the loop of scientific workflows. In *Proceedings of the 6th IEEE (Institute of Electrical and Electronics Engineers) International Conference on e-Science*. Available at https://ieeexplore.ieee.org/document/5693919?arnumber=5693919&tag=1. Accessed June 22, 2020.

Stall, S., L. Yarmey, J. Cutcher-Gershenfeld, B. Hanson, K. Lehnert, B. Nosek, M. Parsons, E. Robinson, and L. Wyborn. 2019. Make scientific data FAIR. *Nature* 570:27–29. doi: 10.1038/d41586-019-01720-7.

Stephan, P. 2012. Research efficiency: Perverse incentives. *Nature* 484:29–31. doi: 10.1038/484029a.

Stephens, T. 2020. Powerful new AI technique detects and classifies galaxies in astronomy image data. *ScienceDaily*. Available at https://phys.org/news/2020-05-powerful-ai-technique-galaxies-astronomy.html. Accessed May 21, 2020.

Stevens, B., C.-H. Moeng, A. S. Ackerman, C. S. Bretherton, A. Chlond, S. de Roode, and P. Zhu. 2005. Evaluation of large-eddy simulations via observations of nocturnal marine stratocumulus. *Monthly Weather Review* 133:1443–1462. doi: 0.1175/MWR2930.1.

Stodden, V. 2020. Cyberinfrastructure shapes scientific outcomes in crucial and largely unrecognized ways. Presentation at the Workshop on Opportunities for Accelerating Scientific Discovery: Realizing the Potential of Advanced and Automated Workflows, March 16–17.

Stoyanovich, J., B. Howe, and H.V. Jagadish. 2020a. Responsible data management. Available at http://www.vldb.org/pvldb/vol13/p3474-stoyanovich.pdf. *Proceedings of the VLDB Endowment* 13(12):3474–3488. doi: 10.14778/3415478.3415570.

Stoyanovich, J., J. J. Van Bavel, and T. West. 2020b. The imperative of interpretable machines. *Nature Machine Intelligence* 2:197–199. doi: 10.1038/s42256-020-0171-8.

Strassler, M., and J. Thaler. 2019. Slow and steady. *Nature Physics* 15(725). doi: 10.1038/s41567-019-0628-z.

Sun, Q., Y. Liu, W. Tian, Y. Guo, and B. Li. 2019. UMDISW: A universal multi-scientific workflow framework for the whole life cycle of scientific data. In *Lecture notes in computer science*, Vol. 11459. Available at https://link.springer.com/chapter/10.1007/978-3-030-32813-9_14. Accessed February 16, 2021.

Sutton, R., and A. Barto. 2018. *Reinforcement learning: An introduction*, 2nd ed. Cambridge, MA: MIT Press.

Szalay, A. S. 2017. From SkyServer to SciServer. *Annals of the American Academy of Political and Social Science* 675(1):202–220. doi: 10.1177/0002716217745816.

Szalay, A. 2019. The era of surveys and the fifth paradigm of science. Available at https://ui.adsabs.harvard.edu/abs/2019AAS...23340001S/abstract. Accessed May 19, 2020.

Szalay, A. 2020. Scalable data aggregation for science. Presentation at the Workshop on Opportunities for Accelerating Scientific Discovery: Realizing the Potential of Advanced and Automated Workflows, March 16–17.

Taylor, I. J., E. Deelman, D. B. Gannon, and M. Shields (eds.). 2007. *Workflows for e-Science: Scientific workflows for grids.* Springer.

Teitelbaum, M. S. 2008. Structural disequilibria in biomedical research. *Science* 321(5859):644–645. doi: 10.1126/science.1160272.

Turner, K., and P. Lambert. 2014. *Workflows for quantitative data analysis in the social sciences.* Los Angeles: Sage. Available at https://core.ac.uk/download/pdf/20323257.pdf. Accessed April 14, 2021.

UKRI (UK Research and Innovation). 2016. Concordat on open research data. Available at https://www.ukri.org/wp-content/uploads/2020/10/UKRI-020920-ConcordatonOpenResearchData.pdf. Accessed April 14, 2021.

UKRI. 2017. UK leads on new 8.5m European scheme to improve access to research data. Available at https://webarchive.nationalarchives.gov.uk/20200923140750/https://stfc.ukri.org/news/uk-leads-on-new-european-scheme. Accessed April 14, 2021.

UKRI. 2020. The UK's research and innovation infrastructure opportunities to grow our capability. Available at https://www.ukri.org/wp-content/uploads/2020/10/UKRI-201020-UKinfrastructure-opportunities-to-grow-our-capacity-FINAL.pdf. Accessed April 14, 2021.

Van der Aalst, W. M. P., and K. van Hee. 2002. *Workflow management: Models, methods, and systems.* Cambridge, MA: MIT Press.

Vidal, R. 2020. Opportunities for accelerating discovery: Mathematical and algorithmic issues. Presentation at the Workshop on Opportunities for Accelerating Scientific Discovery: Realizing the Potential of Advanced and Automated Workflows, March 16–17.

Waibel, G. 2018. Letter to the Community: CDL and Dryad partnership. Available at https://cdlib.org/cdlinfo/2018/05/30/letter-to-the-community-cdl-and-dryad-partnership/. Accessed November 9, 2021.

Wang, L. L., K. Lo, Y. Chandrasekhar, R. Reas, J. Yang, D. Eide, K. Funk, R. M. Kinney, Z. Liu, W. C. Merrill, P. Mooney, D. A. Murdick, D. Rishi, J. Sheehan, Z. Shen, B. Stilson, A. D. Wade, K. Wang, C. Wilhelm, B. Xie, D. A. Raymond, D. S. Weld, O. Etzioni, and S. Kohlmeier. 2020. CORD-19: The COVID-19 Open Research Dataset. *arXiv.* https://arxiv.org/pdf/2004.10706.pdf.

Wasserstein, R. L., and N. A. Lazar. 2016. The ASA statement on p-values: Context, process, and purpose. *American Statistician* 70(2):129–133. doi: 10.1080/00031305.2016.1154108.

Weitzner, D. J. 2020. Challenges of policy-aware data processing. Presentation at the Workshop on Opportunities for Accelerating Scientific Discovery: Realizing the Potential of Advanced and Automated Workflows, March 16–17.

WCRI (World Conferences on Research Integrity). 2019. Hong Kong principles. Available at https://www.wcrif.org/guidance/hong-kong-principles. Accessed February 8, 2021.

Wiebels, K., and D. Moreau. 2021. Leveraging containers for reproducible psychological research. *Advances in Methods and Practices in Psychological Science* 4(2). doi: 10.1177/25152459211017853.

Wilkinson, M., M. Dumontier, I. Aalbersberg, G. Appleton, M. Axton, A. Baak, N. Blomberg, J.-W. Boiten, L. Bonino da Silva Santos, P. E. Bourne, J. Bouwman, A. J. Brookes, T. Clark, M. Crosas, I. Dillo, O. Dumon, S. Edmunds, C. T. Evelo, R. Finkers, A. Gonzalez-Beltran, A. J. G. Gray, P. Groth, C. Goble, J. S. Grethe, J. Heringa, P. A. C. 't Hoen, R. Hooft, T. Kuhn, R. Kok, J. Kok, S. J. Lusher, M. E. Martone, A. Mons, A. L. Packer, B. Persson, P. Rocca-Serra, M. Roos, R. van Schaik, S.-A. Sansone, E. Schultes, T. Sengstag, T. Slater, G. Strawn, M. A. Swertz, M. Thompson, J. van der Lei, E. van Mulligen, J. Velterop, A. Waagmeester, P. Wittenburg, K. Wolstencroft, J. Zhao, and B. Mons. 2016. The FAIR guiding principles for scientific data management and stewardship. *Scientific Data* 3:160018. doi: 10.1038/sdata.2016.18.

Wood, A., M. Altman, A. Bembenek, M. Bun, M. Gaboardi, J. Honaker, K. Nissim, D. R. OBrien, T. Steinke, and S. Vadhan. 2018. Differential privacy: A primer for a non-technical audience. *Vanderbilt Journal of Entertainment & Technology Law* 21(1):209–275. Available at https://salil.seas.harvard.edu/publications/differential-privacy-primer-non-technical-audience. Accessed March 5, 2021.

Woodie, A. 2022. Big growth forecasted for big data. Datanami. January 11. Available at https://www.datanami.com/2022/01/11/big-growth-forecasted-for-big-data/. Accessed April 20, 2022.

Wu, C., G. Wang, S. Hu, Y. Liu, H. Mi, Y. Zhou, Y-K. Guo, and T. Song. 2020. A data driven methodology for social science research with left-behind children as a case study. *PLoS ONE* 15(11):e0242483. doi: 10.1371/journal.pone.0242483.

Yarkoni, T., D. Eckles, J. A. J. Heathers, M. C. Levenstein, P. E. Smaldino, and J. Lane. 2021. Enhancing and accelerating social science via automaton: Challenges and opportunities. *Harvard Data Science Review*. doi: 10.1162/99608f92.df2262f5.

Yuval, J., and P. A. O'Gorman. 2020. Stable machine-learning parameterization of subgrid processes for climate modeling at a range of resolutions. *Nature Communications* 11(1):3295. doi: 10.1038/s41467-020-17142-3.

Appendix A

Workshop Agenda

**Opportunities for Accelerating Scientific Discovery:
Realizing the Potential of Advanced and Automated Workflows**

AGENDA
March 16–17, 2020

March 16, 2020 (Monday)

8:30 am EDT/1:30 pm CET/5:30 am PDT *PART ONE: USE CASES*
Welcome, Overview, and Goals of the Symposium—Daniel Atkins, University of Michigan

8:45 am EDT/1:45 pm CET/5:45 am PDT
Sponsor Perspective—Stuart Feldman, Schmidt Futures

9:00 am EDT/2:00 pm CET/6:00 am PDT
Accelerating Discovery: Case Studies, Requirements, and Progress, Part One
Session Leaders: Ilkay Altintas, San Diego Supercomputer Center, and Tapio Schneider, California Institute of Technology
Panelists: Timothy Cernak, University of Michigan
               Kyle Cranmer, New York University
               Alex Szalay, Johns Hopkins University

10:00 am EDT/3:00 pm CET/7:00 am PDT *BREAK*

10:15 am/3:15 pm CET/7:15 am PDT
Accelerating Discovery: Case Studies, Requirements, and Progress, Part Two
Panelists: Gregory Crane, Tufts University
Tyrel McQueen, Johns Hopkins University
Robert Murphy, Carnegie Mellon University

11:15 am/4:15 pm CET/8:15 am PDT *PART TWO: ENABLING TECHNOLOGIES*
Accelerating Discovery: Mathematical and Algorithmic Issues
Session Leaders: Bradley Malin, Vanderbilt University, Alfred Hero, University of Michigan, Tapio Schneider, California Institute of Technology
Panelists: Rene Vidal, Johns Hopkins University
Rebecca Nugent, Carnegie Mellon University
Victoria Stodden, University of Illinois Urbana-Champaign

12:15 pm/5:15 pm CET/9:15 am PDT *BREAK*

12:30 pm/5:30 pm CET/9:30 am PDT
Perspective from the Office of Science and Technology Policy
Kelvin Droegemeier, Director, Office of Science and Technology Policy

12:40 pm/5:40 pm CET/9:40 am PDT
Accelerating Discovery: Status and Trajectory of Supporting Tools and Systems
Session Leaders: Mercè Crosas, Harvard University, and Shreyas Cholia, Lawrence Berkeley National Laboratory
Panelists: Carole Goble, University of Manchester
Ian Foster, Argonne National Laboratory
Brian Granger, Project Jupyter and Amazon Web Services
Julia Lane, New York University
Peter Beckman, Argonne National Laboratory
Geoffrey Fox, Indiana University

2:40 pm EDT/7:30 pm CET/11:30 am PDT *OPEN DISCUSSION*
(Break as needed)

3:30 pm EDT/8:30 pm CET/12:30 pm PDT *ADJOURN DAY ONE*

March 17, 2020 (Tuesday)

8:30 am EDT/1:30 pm CET/5:30 am PDT **PART THREE: HUMAN AND POLICY ISSUES**

Welcome and Overview of Day Two — Daniel Atkins

8:45 am EDT/1:45 pm CET/5:45 am PDT

Accelerating Discovery: Standards, Governance, and Social Context
Session Leaders: Lara Mangravite, Sage Bionetworks, and Rebecca Lawrence, F1000 Research
Panelists: Raja Mazumder, George Washington University
Beth Plale, National Science Foundation
Timothy Gardner, Riffyn, Inc.
Michael Crusoe, Common Workflow Language

9:45 am EDT/2:45 pm CET/6:45 am PDT **BREAK**

10:00 am EDT/3:00 pm CET/7:00 am PDT

Accelerating Discovery: Developing Supportive Policies, Communities, and Sustainable Funding
Session Leader: Alfred Hero, University of Michigan
Panelists: Jean-Claude Burgelman, European Commission
Dmitri Kusnezov, U.S. Department of Energy
Adam Russell, Defense Advanced Research Projects Agency
Daniel Weitzner, Massachusetts Institute of Technology

11:15 am EDT/4:15 pm CET/8:15 am PDT **PART FOUR: SYNTHESIS AND FUTURE PRIORITIES**

Synthesis of What We Have Heard
Session Leaders: Rebecca Lawrence, F1000 Research, and Tapio Schneider, California Institute of Technology
Panelists: Carole Goble, University of Manchester
Ian Foster, Argonne National Laboratory
Alex Szalay, Johns Hopkins University
Timothy Gardner, Ryffin, Inc.

1:00 pm EDT/6:00 pm CET/10:00 am PDT **ADJOURN PUBLIC WORKSHOP**

Appendix B

Committee Biosketches

DANIEL E. ATKINS (Chair) (NAE) is Emeritus W.K. Kellogg Professor of Information and Professor of Electrical Engineering and Computer Science at the University of Michigan (UM), Ann Arbor. The first phase of his career focused on computer architecture including high-speed arithmetic methods now widely used in modern computers, as well as the design and construction of application-specific experimental computers. The second phase of his career focused on pioneering interdisciplinary research on cyber-enabled distributed knowledge communities including collaboratories and digital libraries applied to both scientific research and education. He has served as dean of the College of Engineering, founding dean of the School of Information, and associate vice president for research at UM, as well as the inaugural director of the Office of Cyberinfrastructure at the National Science Foundation (NSF). He chaired the Blue Ribbon Panel on Research Cyberinfrastructure for the NSF that became an international roadmap for initiatives on cyber-enabled research in the digital age. He has chaired or served on many advisory boards for government, academia, philanthropy, and industry. Professor Atkins is a member of the National Academy of Engineering and Fellow of the AAAS. He earned a Ph.D. in computer science and an M.S. in electrical engineering from the University of Illinois, Urbana-Champaign, and a B.S.E.E. from Bucknell University.

ILKAY ALTINTAS is a research scientist at the University of California San Diego, the chief data science officer of the San Diego Supercomputer Center (SDSC), and a founding Fellow of the Halıcıoğlu Data Science Institute. She is the founding director of the Workflows for Data Science (WorDS) Center of Excellence and the WIFIRE Lab. The WoRDS Center specializes in the development of methods, cyberinfrastructure, and workflows for computational data science and its translation to practical applications. The WIFIRE Lab is focused on artificial intelligence methods for an all-hazards knowledge cyberinfrastructure, becoming a management layer from the data collection to modeling efforts, and has achieved significant success in helping to manage wildfires. Since joining SDSC in 2001, she has been a principal investigator and a technical leader in a wide range of cross-disciplinary projects. With a specialty in scientific workflows, she leads collaborative teams to deliver impactful results through making computational data science work more reusable, programmable, scalable, and reproducible. Her work has been applied to many scientific and societal domains including bioinformatics, geoinformatics, high-energy physics, multiscale biomedical science, smart cities, and smart manufacturing. She is also a popular MOOC instructor in the field of "big" data science and has reached out to more than a million learners across any populated continent. Among the awards she has received are the 2015 IEEE TCSC Award for Excellence in Scalable Computing for Early Career Researchers and the 2017 ACM SIGHPC Emerging Woman Leader in Technical Computing Award. Dr. Altintas received her Ph.D. degree from the University of Amsterdam in the Netherlands, with an emphasis on provenance of workflow-driven collaborative science.

SHREYAS CHOLIA is group leader for the Usable Software Systems Group in the Data Science and Technology department at Lawrence Berkeley National Laboratory (LBNL), focused on usability aspects of computational and data analysis systems. He is particularly interested in how web interfaces and tools can facilitate large-scale scientific computing workflows. He is currently working on various projects that integrate Jupyter Notebooks with distributed and high-performance scientific computing environments. He joined LBNL's Computational Research Division in 2015, having worked for over a decade at the National Energy Research Scientific Computing Center, where he led the science-gateway, web, and grid efforts. Prior to his appointment at LBNL, he was a developer and consultant at IBM. He has a B.A. in computer science and cognitive sciences from Rice University.

MERCÈ CROSAS is the secretary of Open Government for the government of Catalunya. Prior to her current position, Dr. Crosas was the university research data management officer, with Harvard University Information Technology, and chief data science and technology officer at Harvard's Institute for Quantitative Social Science. In the last 10 years, Dr. Crosas has been principal investigator (PI) and co-PI of multiple research grants and collaborations related to data privacy, data provenance, research reproducibility, and data sharing in social science, biomedicine, and astronomy. She is part of numerous committees and working groups focused on research data management, data citation, and data standards, and is a co-author of the FAIR (findable, accessible, interoperable, reusable) data principles as well as the Joint Declaration of Data Citation Principles. Before rejoining Harvard in 2004, Dr. Crosas worked for 6 years in the educational software and biotech industries, initially as a software developer, and subsequently as director of the software development team. She contributed to the development of lab information management systems for single nucleotide polymorphism discovery and genotyping and mass spectrometry. Before that, she spent 6 years at the Harvard-Smithsonian Center for Astrophysics, first as a predoctoral fellow for her Ph.D. in astrophysics from Rice University, and later as a postdoctoral fellow, researcher, and software engineer with the Radioastronomy division. She earned a B.S. in physics from the Universitat de Barcelona, Spain.

ALFRED HERO is the John H. Holland Distinguished University Professor of Electrical Engineering and Computer Science and the R. Jamison and Betty Williams Professor of Engineering at the University of Michigan. His research is on data science and developing theory and algorithms for data collection, analysis, and visualization that use statistical machine learning and distributed optimization. These are being applied to network data analysis, personalized health, multimodality information fusion, data-driven physical simulation, materials science, dynamic social media, and database indexing and retrieval. Dr. Hero has held visiting positions at Massachusetts Institute of Technology, Boston University, Lucent Bell Laboratories (Murray Hill), and Ford Motor Company in addition to the University of Nice, the École Normale Supérieure de Lyon, and Telecom-ParisTech in France. Dr. Hero was president of the Institute of Electrical and Electronics Engineers' (IEEE's) Signal Processing Society (2006–2008) and was on the Board of Directors of IEEE (2009–2011) where he served as director of Division IX (Signals and Applications). He is also a member of the Big Data Special Interest Group of the IEEE Signal Processing Society. Dr. Hero received a B.S. (summa cum laude) from Boston University (1980) and a Ph.D. from Princeton University (1984), both in electrical engineering.

REBECCA LAWRENCE is managing director of F1000 Group. She was responsible for the launch of the open research publishing platform F1000 Research in January 2013, and has subsequently led the initiative behind the recent launch of Wellcome Open Research, Gates Open Research, and many other funder- and institution-based publishing platforms. She is a member of the High-Level Advisory Group for the European Commission's Open Science Policy Platform (OSPP), chairing their work on next-generation indicators and their integrated advice: OSPP-REC. She has been a co-chair of a number of working groups focusing on data and peer review, for organizations including the Research Data Alliance and ORCID. She is also an Advisory Board member for the data policy and standards initiative, FAIRsharing, and for DORA (the San Francisco Declaration on Research Assessment). She has worked in scientific, technical, and medical publishing for almost 20 years for several publishers including Elsevier where she built and ran the Drug Discovery Group. She originally trained and qualified as a pharmacist, and holds a Ph.D. in cardiovascular pharmacology from University of Nottingham.

BRADLEY A. MALIN (NAM) is the vice chair for research and professor of biomedical informatics at Vanderbilt University. He is also a professor of biostatistics, a professor of computer science, and is affiliated faculty in the Center for Biomedical Ethics and Society. He co-directs the Health Data Science Center, the Center for Genetic Privacy and Identity in Community Settings—a National Institutes of Health Center of Excellence in Ethical, Legal, and Social Implications Research, and the Big Biomedical Data Science Ph.D. program. He is also the director of the Health Information Privacy Laboratory, which was established to address the growing need for data privacy research and development for the health information technology sector. Dr. Malin's research is in big health data analytics and the infrastructure necessary to support such investigations. He has made specific contributions to a number of health-related areas, including distributed data processing methods for medical record linkage and predictive modeling, intelligent auditing technologies to protect electronic medical records from misuse in the context of primary care, and algorithms to formally anonymize patient information disseminated for secondary research purposes. He is an elected fellow of the National Academy of Medicine and American College of Medical Informatics and was honored as a recipient of the Presidential Early Career Award for Scientists and Engineers. Dr. Malin completed his education at Carnegie Mellon University, where he received a bachelor's degree in biological sciences, a master's in machine learning, a master's in public policy and management, and a doctorate in computer science (with a focus on databases and software systems).

LARA MANGRAVITE is president of Sage Bionetworks. This organization is focused on the development and implementation of practices for large-scale collaborative biomedical research. Sage Bionetworks' work is centered on new approaches to scientific process that use open systems to enable community-based research regarding complex biomedical problems. Previously, Dr. Mangravite served as director of the Systems Biology research group at Sage Bionetworks where she focused on the application of collaborative approaches to advance understanding of disease biology and treatment outcomes at a systems level with the overriding goal of improving clinical care. Dr. Mangravite obtained a B.S. in physics from Pennsylvania State University and a Ph.D. in pharmaceutical chemistry from the University of California, San Francisco. She completed a postdoctoral fellowship in cardiovascular pharmacogenomics at the Children's Hospital Oakland Research Institute.

TAPIO SCHNEIDER is Theodore Y. Wu Professor of Environmental Science and Engineering at Caltech and Senior Research Scientist at the Jet Propulsion Laboratory. His research is focused on understanding atmosphere dynamics on Earth and other planets; turbulence in atmosphere and oceans; and climate change and climate modeling. Previously, Dr. Schneider served as professor of climate dynamics at Swiss Federal Institute of Technology Zurich from 2013 to 2016, and associate research scientist at New York University's Courant Institute of Mathematical Sciences from 2000 to 2002. Dr. Schneider received his M.Sc. (1997) and Ph.D. (2001) in atmospheric and oceanic sciences from Princeton University. He was a visiting graduate student (Physics) at the University of Washington, Seattle, from 1994 to 1995, and studied mathematics and physics (Vordiplom 1993) at Albert-Ludwigs-Universität Freiburg, Germany.